A First Course in Electrode Process
2nd Edition

A First Course in Electrode Processes
2nd Edition

Derek Pletcher
School of Chemistry, University of Southampton, Southampton, UK

RSCPublishing

ISBN: 978-1-85755-893-0

A catalogue record for this book is available from the British Library

Published by The Royal Society of Chemistry,
Thomas Graham House, Science Park, Milton Road,
Cambridge CB4 0WF, UK

Registered Charity Number 207890

For further information see our web site at www.rsc.org

Preface

I have been a teacher of electrochemistry for over 40 years. This period has seen many advances. There has been a growth in the applications of electrolysis in industry. This has been accompanied by improvements in both electrolysis cell design and the availability of high-performance cell components. In parallel, there have been massive improvements in the instrumentation used in the laboratory. In the mid-1960s it was unusual to finish an experiment without the aid of an electronics technician while now, the instrumentation is reliable and also allows much improved presentation of results as well as routine, automated analysis of the data. These developments have led to a greater understanding of electrode reactions and made possible the use of electrochemical techniques to study systems of interest to medicine, material science and engineering as well as many areas of chemistry. These advances inevitably impact the way we teach the subject, whether to undergraduates, postgraduates or mature scientists/engineers. Moreover, electrochemistry is recognized to be important to the future. Perhaps the greatest problems facing society are to establish stable energy supplies and to minimize our impact on the environment; electrochemistry can contribute substantially to the solution of both.

There are many textbooks on electrochemistry but few that I feel able to recommend to those starting to learn the subject. After their initial contact with textbooks, most beginners conclude that electrochemistry is a subject dependent on complex theories and abundant equations and fail to see the concepts that underpin the understanding and treatment of electrode reactions. In consequence, it has been my aim to produce a book that can be read by all those with some background in chemistry

A First Course in Electrode Processes, 2nd Edition
By Derek Pletcher
© Derek Pletcher 2009
Published by the Royal Society of Chemistry, www.rsc.org

and/or used as the basis of an introductory course on electrode processes. I have sought to emphasize the concepts and understanding of
electrode reactions, limiting the presentation to essential equations,
usually without the detailed mathematical derivations that can be found
elsewhere. Wherever possible, the text also includes discussion of illus-
trative applications. The final chapter contains several problems with
detailed answers. I also believe that courses should include practicals
and/or demonstrations. It is not difficult to set up inexpensive experiments with simple electronics, well-designed cells and systems such as
the electrochemistry of $Fe(CN)_6^{4-}$ in $1\,M\,KCl$, $Cu(\textrm{II})$ in $1\,M\,NaCl$,
nitrobenzene in acetonitrile, Pt in $1\,M\,H_2SO_4$ or Ni in $1\,M\,NaOH$.

Electrochemists have the advantages that it is straightforward to
demonstrate that the topic is 'relevant' and interacts with many areas of
science and technology. All students in chemistry, chemical engineering
and material science would benefit from some exposure to electrochemistry. The aim should be to infiltrate some electrochemistry into
more undergraduate and postgraduate courses and ensure that gradu-
ates understand the possibilities opened up by electrolysis and electrochemical experiments.

This is the second edition of *A First Course of Electrode Processes*. It
retains the same structure as the first edition, published in 1991, but all
chapters have been rewritten and, in some cases, the balance has been
changed. While the last two chapters before the problems are still intended
to illustrate the application of the earlier material to technology, the topics
have changed. Now the chapters are devoted to fuel cells and the applications of electrochemistry to improving the environment, reflecting
major thrusts for the future. In addition, a new chapter (Chapter 6) on
experimental factors has been introduced.

As always, I must acknowledge the great contributions of my wife,
Gill, to this book. Her encouragement and support is critical to all my
activities. She also has a particular role in this book; she has used her
experience of editing electrochemical manuscripts to remove many
grammatical errors and misuse of symbols and terminology from the
text.

Derek Pletcher

Contents

A First Course in Electrode Processes, 2nd Edition
By Derek Pletcher
© Derek Pletcher 2009
Published by the Royal Society of Chemistry, www.rsc.org

List of Symbols

This list includes the definition and common/typical units of symbols used repetitively throughout the text.

a	activity of a species in solution	–
	or radius of a microdisc electrode	μm
A	electrode area	cm^2
c	concentration of a species in solution	M or mol cm^{-3}
C	capacitance	$F\,cm^{-2}$
D	diffusion coefficient	$cm^2\,s^{-1}$
E	experimental potential vs reference electrode	V
E_e	equilibrium potential for a couple vs reference electrode	V
E_e^o	formal potential for a couple vs reference electrode – the equilibrium potential when $c_O = c_R$	V
E_{cell}	cell voltage, $i.e.$ $E_c - E_a$	V
$E_{1/2}$	half-wave potential, $i.e.$ the potential where $I = 0.5 I_L$	V
f	frequency of rotation or AC signal	s^{-1}
F	the Faraday constant	$96485\,C\,mol^{-1}$
ΔG	Gibbs free energy change associated with a reaction	$J\,mol^{-1}$
I	current	A
I_L	limiting current	A
I_o	exchange current	A
j	experimental current density	$A\,cm^{-2}$
j_L	limiting current density	$A\,cm^{-2}$
j_o	exchange current density	$A\,cm^{-2}$

A First Course in Electrode Processes, 2nd Edition
By Derek Pletcher
© Derek Pletcher 2009
Published by the Royal Society of Chemistry, www.rsc.org

J	flux of material	$\mathrm{mol\,cm^{-2}\,s^{-1}}$
k	rate constant for a chemical reaction	$\mathrm{s^{-1}}$
	or the gas constant per molecule	$\mathrm{J\,K^{-1}}$
k_s	standard rate constant for an electron-transfer reaction – the rate constant for both oxidation and reduction at the formal potential	$\mathrm{cm\,s^{-1}}$
k_a	rate constant for an anodic reaction at any potential	$\mathrm{cm\,s^{-1}}$
k_c	rate constant for a cathodic reaction at any potential	$\mathrm{cm\,s^{-1}}$
k_m	mass transfer coefficient	$\mathrm{cm\,s^{-1}}$
K	equilibrium constant	–
m	moles of reactant	mol
M	molecular weight	$\mathrm{g\,mol^{-1}}$
n	number of electrons involved in electrode reaction	–
N	collection efficiency of a RRDE (rotating ring disc electrode)	–
q	charge density on a surface	$\mathrm{C\,cm^{-2}}$
Q	charge	C
r	radius of an ion	cm
R	resistance	ohms
	or gas constant	$\mathrm{J\,K^{-1}\,mol^{-1}}$
R_u	uncompensated resistance between working electrode and reference electrode probe	ohms
S	separation of two electrodes	cm
t	time from commencement of an experiment	s
T	temperature	K
v	electrolyte flow rate	$\mathrm{cm\,s^{-1}}$
V	volume	$\mathrm{cm^3}$
V_{cell}	cell/battery voltage	V
x	distance perpendicular to the electrode surface	cm
z	charge on an ion	–
Z	impedance	ohm
α	transfer coefficient	–
δ	thickness of mass transfer layer at an electrode surface	cm
ε	dielectric constant of a medium	–
ϕ	fractional charge (current) efficiency	–
ϕ_M	absolute potential on the electrode surface	V
ϕ_S	absolute potential in the solution adjacent to the electrode surface	V
γ	activity coefficient	–
	or surface tension	$\mathrm{dyne\,cm^{-1}}$
η	overpotential	V
	or viscosity	poise

κ	conductivity	$S\,cm^{-1}$
λ	reorganization energy	$J\,mol^{-1}$
v	potential scan rate	$V\,s^{-1}$
	or kinematic viscosity	$cm^2\,s^{-1}$
θ	fractional coverage by an adsorbate	–
ρ	density	$g\,cm^{-3}$
	or charge density in solution	$C\,cm^{-3}$
$\tau_{1/2}$	half-life of an intermediate	s
ω	rotation rate of a RDE (rotating disc electrode)	$radians\,s^{-1}$

Subscripts: 'a' and 'c' refer to anodic and cathodic reactions, respectively, 'i' to a species and 'O' and 'R' to the oxidized and reduced species, respectively. Prefixes to units: k = kilo (10^3), m = milli (10^{-3}), μ = micro (10^{-6}), n = nano (10^{-9}) and p = pico (10^{-12}).

CONVENTIONS

Anodic (oxidation) currents are positive and cathodic (reduction) currents are negative.

Making the potential more positive will increase the driving force for oxidation, while making it more negative will increase the driving force for reduction.

CHAPTER 1

An Introduction to Electrode Reactions

1.1 INTRODUCTION

Electrode reactions are a class of chemical reactions that involve the transfer of a charged species across an interface, most commonly an electron between a conducting solid and an adjacent solution phase. The electron is negatively charged and the transfer is driven by a voltage gradient across the interface created by applying a potential between the conducting solid (an electrode) and a second electrode.

Electrode reactions are central to numerous diverse technologies (Table 1.1), many of which are contributing to a healthier and cleaner environment and their importance will grow as the pressure increases to provide clean water, reduce CO_2 emissions and to use resources more efficiently. They also underlie several natural phenomena. In addition, laboratory electrochemical techniques provide approaches to understanding a broad range of chemical systems related to synthesis, biology, materials, component fabrication, energy conservation and engineering.

A simple type of electrode reaction involves only electron transfer between an inert metal electrode and an ion or molecule in solution. At an anode, the electron passes from the solution phase to the electrode and the electroactive species in solution is oxidized, *e.g.*:

$$Fe^{2+} - e^- \longrightarrow Fe^{3+} \qquad (1.1)$$

Conversely, at a cathode the electron passes in the opposite direction so that the species in solution is reduced, *e.g.*:

$$Cr^{3+} + e^- \longrightarrow Cr^{2+} \qquad (1.2)$$

A First Course in Electrode Processes, 2nd Edition
By Derek Pletcher
© Derek Pletcher 2009
Published by the Royal Society of Chemistry, www.rsc.org

Table 1.1 Some applications of electrode reactions.

Manufacture of both inorganic and organic chemicals
Extraction, refining and recycling of metals
Primary and secondary batteries
Fuel cells
Deposition of coatings – metals, oxides, semiconductors and paints
Mechanical component manufacture – electrochemical machining and electroforming
Manufacture of electronic components
Purification and extraction of chemicals
Process stream conditioning and recycling
Effluent treatment
Purification of water supplies
Waste disposal
Minimizing corrosion
Laboratory analytical procedures
Sensors – in hospitals and environmental monitoring

Clearly, such electron-transfer reactions are going to be possible only when the electroactive species is within molecular distances of the electrode surface (maybe, even a bond between the reactant and the surface is necessary). Hence, even with these simple electrode reactions, the continuous conversion of reactant into product must be a multistep process involving the supply of reactant to the surface and the removal of the product in addition to the electron-transfer event. The overall conversion of reactant, A, into product, B, must occur in a minimum of three steps:

$$A_{soln} \xrightarrow{\text{mass transport}} A_{electrode} \qquad (1.3)$$

$$A_{electrode} \xrightarrow{\text{electron transfer}} B_{electrode} \qquad (1.4)$$

$$B_{electrode} \xrightarrow{\text{mass transfer}} B_{solution} \qquad (1.5)$$

The rate of conversion of reactant in the bulk solution into product in the bulk solution must always be determined by the slowest of these three steps. As a result, to develop an understanding of electrode reactions, one must first discuss the physical chemistry of both electron transfer and mass transfer as well as explaining the way in which the two processes interact. These themes will be developed later in this chapter.

First, however, it is important to emphasize that most electrode reactions are more complex. This can be seen from the examples of both cathode and anode reactions in Table 1.2.

Table 1.2 Typical electrode reactions of interest in the laboratory and in electrochemical technology – note the reactions go from single electron transfers to multistep processes involving several electrons and the formation/cleavage of bonds.

Cathode reactions

$$2H^+ + 2e^- \rightarrow H_2 \tag{1.6}$$

$$2H_2O + 2e^- \rightarrow H_2 + 2OH^- \tag{1.7}$$

$$Fe(CN)_6^{3-} + e^- \rightarrow Fe(CN)_6^{4-} \tag{1.8}$$

$$Cu^{2+} + 2e^- \rightarrow Cu \tag{1.9}$$

$$PbO_2 + SO_4^{2-} + 4H^+ + 2e^- \rightarrow PbSO_4 + 2H_2O \tag{1.10}$$

$$2CH_2{=}CHCN + 2H_2O + 2e^- \rightarrow (CH_2CH_2CN)_2 + 2OH^- \tag{1.11}$$

Anode reactions

$$2H_2O - 4e^- \rightarrow O_2 + 4H^+ \tag{1.12}$$

$$H_2 - 2e^- \rightarrow 2H^+ \tag{1.13}$$

$$Ce^{3+} - e^- \rightarrow Ce^{4+} \tag{1.14}$$

$$Fe - 2e^- \rightarrow Fe^{2+} \tag{1.15}$$

$$2Al + 3H_2O - 6e^- \rightarrow Al_2O_3 + 6H^+ \tag{1.16}$$

$$2Cl^- - 2e^- \rightarrow Cl_2 \tag{1.17}$$

$$Ni(OH)_2 + OH^- - e^- \rightarrow NiO(OH) + H_2O \tag{1.18}$$

$$C_6H_5CH_3 + 2CH_3OH - 4e^- \rightarrow C_6H_5CH(OCH_3)_2 + 4H^+ \tag{1.19}$$

$$CH_3OH + H_2O - 6e^- \rightarrow CO_2 + 6H^+ \tag{1.20}$$

Inspection of these reactions shows that several other fundamental steps, in addition to electron transfer and mass transfer, may be important. These include:

1. *Adsorption*: The simplest model of electron transfer would envisage that the electron transfers as the reactant 'bounces' on the electrode surface. For many electrode reactions, it is more likely that there is a chemical interaction between the reactant (and/or product) and the electrode surface, *i.e.* adsorption occurs. The interaction may be of varied strength and either electrostatic (between an ion or dipole in solution and the charged electrode surface) or a covalent bond. Hydrogen evolution, Reactions (1.6) or (1.7), is the classical example of a reaction that involves adsorbed hydrogen atoms as intermediates.

2. *Chemical reactions*: Commonly, the initial product of electron transfer is not stable in the environment of the electrode reaction and the conversion of reactant into product involves chemical reactions. Thus, Reactions (1.6), (1.7), (1.10–1.13) and (1.16–1.20) all involve the cleavage or formation of chemical bond(s). The chemistry may be either homogeneous in solution as an intermediate moves away from the electrode or heterogeneous with the intermediate adsorbed on the electrode surface. An example of the former is Reaction (1.19) which involves cationic intermediates (the cation radical of toluene and later the benzyl cation) and these react with the nucleophilic solvent, methanol. In contrast, in Reaction (1.6) the H–H bond is formed by coupling of adsorbed H atoms.

3. *Phase formation*: Some electrode reactions involve the formation of a new phase, a metal lattice in the case of Reaction (1.9), an oxide layer in Reaction (1.16), the interconversion of two solid phases in Reactions (1.10) and (1.18) or the formation of a gas in Reactions (1.6), (1.7) and (1.12). The characteristics of the electrode reaction will depend on the properties of the new phase (*e.g.* is it conducting?) and by the need for nucleation of the new phase and thereafter by the way it grows.

4. *Multiple electron transfer*: Many of the reactions in Table 1.2 involve the transfer of more than one electron. In most cases, the electrons will not be transferred consecutively. Rather the electrode reactions are a cascade of events with the electron transfers separated by other steps such as chemical reactions.

Clearly, a complete description of an electrode reaction may need to include a statement of the number and type of fundamental steps, the order in which they occur and the thermodynamics and kinetics of each step. While this is seldom possible for complex reactions, we shall wish to identify the key steps in the overall sequence, especially the rate-determining step and how the rate may be changed by modification of the electrolysis conditions (choice of electrode material, solution conditions, temperature, *etc.*). In particular, when discussing experiments, emphasis will be placed on experimental tests or data treatments that allow the identification of key steps.

Table 1.2 also illustrates the wide range of chemistry possible at an electrode. Notably, the reactant/product may be:

- inorganic, organic or a biomolecule;
- a solid (even the electrode itself), a gas, the solvent or a dissolved species;
- at either anode or cathode, the reactant may be a cation, an anion or a neutral species.

Moreover, the electrode may be a metal, another material with metal-like conductivity (a carbon or conducting polymer) or a semiconductor. It may also be a bulk material, a coating on an inert, conducting substrate or a three-dimensional structure such as a foam or a gas diffusion electrode. It is also interesting to note the range of currents and current densities that may be met. In the laboratory experiment, the currents measured may commonly range from 1 nA to 1 A while in the applications of Table 1.1, the currents of interest may range from 1 μA to 10^6 A. Overall, current densities of interest may range from 1 pA cm^{-2} to 1 A cm^{-2}.

1.2 ELECTROLYSIS CELLS

Figure 1.1 shows a sketch of a typical electrolysis cell, in fact a modern membrane cell for the manufacture of chlorine and sodium hydroxide. The electrode reactions are (1.7) and (1.17) so that the net chemical change in the cell is:

$$2Na^+ + 2Cl^- + 2H_2O \longrightarrow Cl_2 + H_2 + 2OH^- + 2Na^+ \qquad (1.21)$$

Many of the characteristics of an electrolysis cell may be understood by noting the need for charge transport throughout a complete circuit consisting of the cell and external connections between the two electrodes. Moreover, it is necessary to avoid charge accumulation at

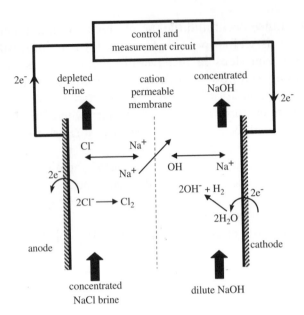

Figure 1.1 Sketch of a modern membrane cell for the manufacture of chlorine and sodium hydroxide, drawn to stress the various electron and transport processes within the cell.

any point in the system. For each molecule of chlorine formed, two chloride ions will be oxidized and two electrons will cross the anolyte/anode interface. To maintain charge neutrality throughout the system, several consequences follow:

1. The negative charge lost from the solution at the anode must be replaced by the transfer of two electrons from the cathode to the solution. Hence, in terms of electrons, the amount of oxidation at the anode and reduction at the cathode must be the same – the formation of 1 mole of chlorine at the anode will be accompanied by the formation of 1 mole of hydrogen and 2 moles of hydroxide at the cathode.

2. To maintain charge neutrality throughout the solution, ions must move between the electrodes – anions towards the anode and/or cations towards the cathode. Since the only requirement is charge balance, it is not necessarily the same ion moving throughout the interelectrode gap. In the cell shown in Figure 1.1, the charge will be carried by sodium and chloride ions within the anolyte and sodium and hydroxide ions in the catholyte (the fraction carried by each ion depending on their transport numbers, see Chapter 2). The cell chemistry, however, is dependent on a membrane between

anolyte and catholyte that allows only the transport of sodium ions.
3. To avoid accumulation of charge on the electrodes, two electrons must pass through the external circuit from anode to cathode.

This last statement leads to several consequences. Firstly, the number of electrons/second passing through the external circuit must be the same as the number of electrons/second crossing the anode/solution and cathode/solution interfaces. In consequence, measurement of the number of electrons/second through the external circuit (*i.e.* the current, I, when the flux of electrons is converted into amperes) gives a direct measure of the rate of chemical change at the electrode surfaces. In electrochemistry, the current or current density, j (current per unit area of electrode surface), always provides a trivial way of monitoring the instantaneous rate of chemical change within the cell. Secondly, the total amount of chemical change within a period of time must be related to the total number of electrons passed through the external circuit during this time period. In other words, it is necessary to integrate the current with respect to time, t, to obtain the total charge passed, Q, and then to apply Faraday's law:

$$Q = \int I dt = mnF \qquad (1.22)$$

where m is the number of moles of reactant consumed or product formed, n is the number of electrons required to convert the reactant into product and F is the Faraday constant ($96485\,C\,mol^{-1}$). Notably, Faraday's law is really just a statement of mass balance. In the reaction:

$$O + ne^- \longrightarrow R \qquad (1.23)$$

the conversion of 1 mole of O into 1 mole of R requires n moles of electrons and the Faraday constant is the charge on 1 mole of electrons (in fact, the Avogadro constant, 6.02×10^{23} electrons per mole, multiplied by the charge on a single electron, 1.60×10^{-19} C). It might also be noted that the rate of conversion of reactant into product at the electrode surface is proportional to the current density and a current density of $0.1\,A\,cm^{-2}$ corresponds to the formation of $1.8\,mmol\,cm^{-2}\,h^{-1}$ for a $2e^-$ reaction.

So far it has been assumed that only one reaction occurs at each electrode. In reality, there are often competing reactions. For example, in the cell of Figure 1.1, the formation of chlorine, Reaction (1.17) is actually accompanied by a small amount of oxygen evolution, Reaction (1.12). Then the current efficiency, ϕ, must be defined:

$$\phi = \frac{\text{charge consumed in reaction of interest}}{\text{total charge passed}} \tag{1.24}$$

The current for the reaction of interest is ϕI and the charge consumed by this reaction is ϕQ.

As noted above, the overall chemical reaction in the cell of Figure 1.1 is given by Equation (1.21). The Gibbs free energy change, ΔG, associated with the reaction may be estimated from tables of thermodynamic data; as written, the free energy change is 419 kJ mol^{-1} of chlorine. Clearly, the reaction is very unfavourable, in accord with our knowledge that a solution of sodium chloride is highly stable. As a result, for the electrolysis to be driven, energy must be applied by applying a voltage between the two electrodes. The magnitude of the essential equilibrium cell voltage, $E_{\text{cell}}^{\text{e}}$, may be calculated from:

$$\Delta G = -nFE_{\text{cell}}^{\text{e}} \tag{1.25}$$

Such a calculation leads to an equilibrium cell voltage of $-2.17\,\text{V}$. This cell voltage could also be calculated from:

$$E_{\text{cell}}^{\text{e}} = E_{\text{c}}^{\text{e}} - E_{\text{a}}^{\text{e}} \tag{1.26}$$

where the equilibrium potentials of the cathode and anode, E_{c}^{e} and E_{a}^{e}, respectively, may be calculated from the Nernst equation (see the next section).

In practice, applying the equilibrium cell voltage would not be sufficient for electrolysis to occur. To achieve electrolysis at a significant rate, it is necessary to apply an overpotential, η, to each electrode to increase the rate of electron transfer (again, see the next section) and also apply additional voltages to drive the ions through the electrolytes and membrane. These voltages will depend on the cell current and the resistance of the components and will be equal to IR_{soln} and IR_{mem}, respectively. Overall, the cell voltage required to have an electrolysis current, I, will be given by:

$$E_{\text{cell}} = E_{\text{cell}}^{\text{e}} - |\eta_{\text{c}}| - |\eta_{\text{a}}| - |IR_{\text{soln}}| - |IR_{\text{mem}}| \tag{1.27}$$

where the modulus symbols are used to ensure that the last four terms all make E_{cell} a larger more negative voltage, thereby increasing the energy consumption for the electrolysis. The overpotentials and IR terms are inefficiencies and the associated energy ends up as heat. In technology, the cell design and choice of materials will seek to minimize these terms.

Notably, in laboratory experiments to investigate mechanism and kinetics it is desirable that the cell response is determined by a single electrode (the working electrode) and that *IR* terms do not distort the measurements. This can be achieved in a two-electrode cell provided the second electrode does not influence the response in any way; this is normally when the second electrode is a reliable reference electrode and the cell current is low; the low cell current also ensures that the *IR* drops are insignificant. More commonly, however, a three-electrode cell and a potentiostat are employed. In this situation the current is passed between the working and a counter electrode (usually much larger than the working electrode) and the potential of the working electrode is controlled *versus* a reference electrode. The potentiostat is a feedback circuit based on an operational amplifier that ensures that the current through the reference electrode is effectively zero. If the current between working and reference electrodes is zero, then the *IR* drop must also be zero and the potential of the working electrode can be measured directly with a digital voltmeter and the value will not be affected by *IR* drop. The design of laboratory cells is discussed in more detail in Chapter 6.

1.3 SIMPLE ELECTRON-TRANSFER REACTIONS

In this section, we consider the thermodynamics and kinetics of a simple electron-transfer reaction:

$$O + ne^- \rightleftharpoons R \tag{1.28}$$

Throughout we shall consider a cell with an inert working electrode and a reference electrode in a solution containing low concentrations of both O and R, c_O and c_R, respectively, and a high concentration of an inert electrolyte (most importantly, to give the solution conductivity). The only electrode reaction that can occur at the electrode, at least in the potential range under consideration, is the interconversion of O and R.

1.3.1 Equilibrium Potential

Initially it is helpful to specify the situation at equilibrium. The simple experiment set out in Figure 1.2 can be used as illustration. The figure shows the two electrodes dipping into the solution specified above and the two electrodes are connected by a circuit containing a high

Figure 1.2 Simple set-up for determining the equilibrium potential for the couple O/R in solution. The cell consists of a vitreous carbon disc electrode and a saturated calomel reference electrode in a solution containing O and R and 1 M KCl.

impedance digital voltmeter; this allows the potential difference between the electrodes to be monitored but prevents the passage of current through the external circuit.

In the absence of a current, the concentrations of O and R cannot change and the working electrode will take up the equilibrium potential for the couple in solution, E_e. Hence, the equilibrium potential can be read with the digital voltmeter. It could also be calculated from the Nernst equation:

$$E_e = E_e^o + \frac{2.3RT}{nF} \log \frac{c_O}{c_R} \tag{1.29}$$

where E_e^o is the formal potential for the couple O/R; it is clearly the equilibrium potential when the concentrations of O and R are equal. The formal potential reflects the ease of addition of an electron to O and removal of an electron from R and is determined by the chemistry of O and R in the particular solution under investigation. For example, the addition of a complexing agent that stabilizes O more than R will make O more difficult to reduce and lead to a negative shift in the formal potential.

Although seldom carried out in practical electrochemistry, the thermodynamic equation, (1.29), should strictly be written in terms of activities. For the general electrode reaction:

$$p\text{P} + q\text{Q} + ne^- \rightleftharpoons x\text{X} + y\text{Y} \tag{1.30}$$

the Nernst equation in the more precise form would be written:

$$E_e = E_e^\circ + \frac{2.3RT}{nF} \log \frac{(a_\text{P})^p (a_\text{Q})^q}{(a_\text{X})^x (a_\text{Y})^y} \tag{1.31}$$

where E_e° is the standard potential (the equilibrium potential when all the reactants and products are in their standard states). While it is wise always to consider the approximations involved in using concentrations rather than activities (*e.g.* it would be unwise if the solution contained a high concentration of electroactive species and no excess of electrolyte), the discussion throughout this book will use 'concentration' and the 'formal potential'. This is partly because for dilute solutions of reactant and product in the presence of a large excess of inert electrolyte the activity coefficients of O and R are likely to be similar and therefore cancel in the Nernst equation. In addition, notably, for the reaction:

$$\text{M}^{n+} + ne^- \rightleftharpoons \text{M} \tag{1.32}$$

the Nernst equation may be written:

$$E_e = E_e^\circ + \frac{2.3RT}{nF} \log c_{\text{M}^{n+}} \tag{1.33}$$

since the activity of a metal is 1 (the measured potential is independent of the amount of the metal used). Also by definition, the activities of elements and gases at a pressure of 1 atmosphere are 1, often allowing further simplification of Nernst equations.

Standard electrode potentials are readily available in textbooks and handbooks of physical chemistry and are the starting point for precise thermodynamic calculations. More often, however, they are used for qualitative assessments of cell potentials and then they should be

Table 1.3 Some typical standard potentials in aqueous solution, pH 0. Much
more complete tables may be found in the *Handbook of Chemistry
and Physics*, CRC Press, Boca Raton, Florida.

Reaction	E_e^o vs SHE/V
$S_2O_8^{2-} + 2e^- \rightleftarrows 2SO_4^{2-}$	+2.00
$Ag^{2+} + e^- \rightleftarrows Ag^+$	+1.99
$MnO_4^- + 8H^+ + 5e^- \rightleftarrows Mn^{2+} + 4H_2O$	+1.49
$Ce^{4+} + e^- \rightleftarrows Ce^{3+}$	+1.44
$Cl_2 + 2e^- \rightleftarrows 2Cl^-$	+1.36
$Cr_2O_7^{2-} + 14H^+ + 6e^- \rightleftarrows 2Cr^{3+} + 7H_2O$	+1.33
$O_2 + 4H^+ + 4e^- \rightleftarrows 2H_2O$	+1.23
$Br_2 + 2e^- \rightleftarrows 2Br^-$	+1.06
$Ag^+ + e^- \rightleftarrows Ag$	+0.80
$Fe(CN)_6^{3-} + e^- \rightleftarrows Fe(CN)_6^{4-}$	+0.69
$Cu^{2+} + 2e^- \rightleftarrows Cu$	+0.34
$CO_2 + 6H^+ + 6e^- \rightleftarrows CH_3OH + H_2O$	+0.03
$2H^+ + 2e^- \rightleftarrows H_2$	0.00
$Cd^{2+} + 2e^- \rightleftarrows Cd$	−0.40
$Fe^{2+} + 2e^- \rightleftarrows Fe$	−0.41
$Zn^{2+} + 2e^- \rightleftarrows Zn$	−0.76
$Al^{3+} + 3e^- \rightleftarrows Al$	−1.71
$Mg^{2+} + 2e^- \rightleftarrows Mg$	−2.37
$Na^+ + e^- \rightleftarrows Na$	−2.71
$Li^+ + e^- \rightleftarrows Li$	−3.04

regarded only as guidelines as they apply to standard state conditions
and not necessarily to the experimental conditions. Table 1.3 presents a
short list of standard potentials; in accordance with general practice,
they are quoted *versus* the standard hydrogen electrode (SHE). This is
seldom the reference electrode in the laboratory but conversion into the
experimental reference electrode is straightforward (Chapter 6).

Returning to the experiment of Figure 1.2, no current is passing
through the cell and therefore the composition of the solution cannot
change. It has been noted that the inert electrode will take up the
equilibrium potential for the O/R couple with the particular solution
concentrations of O and R. Now, it needs to be recognized that,
in common with other chemical systems at equilibrium, a dynamic
situation will prevail at the electrode surface – both reduction of O to R
and oxidation of R to O will be occurring but these reactions will take
place at the same rate. Hence, in terms of current densities, one can write:

$$j = j_a + j_c = 0 \tag{1.34}$$

where the partial anodic and cathodic current densities, j_a and j_c,
respectively, have opposite signs since during oxidation and reduction at

the electrode the electrons pass in the opposite direction across the electrode/solution interface. By convention, anodic currents are positive and cathodic currents are negative. The magnitude of these partial current densities at equilibrium turns out to be a useful kinetic parameter for the O/R couple and hence it is given a name 'the exchange current density', j_o:

$$j_a = -j_c = j_o \tag{1.35}$$

The exchange current density is a measure of the electron-transfer activity at the equilibrium potential. A large value indicates that there is extensive oxidation and reduction occurring while a low value reflects only a small amount. The value also depends on the concentrations of O and R since both oxidation and reduction are first-order reactions in their reactant.

1.3.2 Other Potentials

Suppose that a potential positive to the equilibrium potential is imposed on the inert electrode in the solution containing O and R and an excess of electrolyte. The system will now seek to move to a new equilibrium where the concentrations of O and R are those demanded by the Nernst equation for this potential. At a potential of $E + \Delta E$ the equality of the Nernst equation, Equation (1.29), can be met only by the ratio c_O/c_R increasing. This requires conversion of R into O so that it must be concluded that at a potential positive to the equilibrium potential an anodic current is to be expected. The same argument will lead to the conclusion that for a potential $E - \Delta E$, *i.e.* negative to the equilibrium potential, a net cathodic current will be observed.

The above is only a thermodynamic prediction. The next step is therefore to consider the rate (or current density) at which O and R are interconverted at any potential. Following classical kinetics, it is to be expected that the rates of oxidation and reduction will be the product of a rate constant and the concentration of a reactant at the site of electron transfer, *i.e.* the electrode surface (denoted by the subscript $x = 0$). Hence we can write:

$$\text{Rate of oxidation} = k_a (c_R)_{x=0} \tag{1.36}$$

$$\text{Rate of reduction} = k_c (c_O)_{x=0} \tag{1.37}$$

Notably, the rates of conversion on the surface will have the units moles $\text{cm}^{-2}\text{s}^{-1}$ and therefore the rate constants have the units cm s^{-1}. The rate of the heterogeneous electron transfer would be expected to depend on the potential gradient at the surface driving the movement of the negatively charged electron between the electrode and solution phases. The gradient, in turn, depends on the potential of the electrode (Chapters 3 and 4). It is, however, not straightforward to predict theoretically the way in which the rate constants vary with potential. On the other hand, it is found experimentally that the relationships are generally of the form:

$$k_a = k_a^o \exp \frac{\alpha_a nFE}{RT} \tag{1.38}$$

$$k_c = k_c^o \exp -\frac{\alpha_c nFE}{RT} \tag{1.39}$$

where α_a and α_c are known as the anodic and cathodic transfer co-efficients, respectively. For simple electron-transfer reactions $\alpha_a + \alpha_c = 1$ and commonly both have a value close to 0.5. The interpretation of the transfer coefficients and the form of the potential dependence are discussed in more detail in Chapter 3. Moreover, for now, it will also be assumed the concentrations of O and R at the electrode surface are the same as those in the bulk solution (*i.e.* those prepared for the experiment). This is equivalent to limiting the discussion to conditions where the surface reactions occur at a low rate. Other situations are discussed in Section 1.5 after mass transport has been introduced in Section 1.4. The rates of oxidation and reduction may be converted into current densities simply be multiplying by nF (which will convert the dimensions from moles $cm^{-2} s^{-1}$ into A cm^{-2}). Hence, using Equations (1.36–1.39) and remembering the sign convention for current density, an expression for the current density at any potential may be written:

$$j = j_a + j_c$$
$$= nFk_a^o c_R \exp \frac{\alpha_a nFE}{RT} - nFk_c^o c_O \exp -\frac{\alpha_c nFE}{RT} \tag{1.40}$$

This equation describes the current density as a function of the electrode potential measured *versus* a reference electrode. A simpler and more useful form is obtained by introducing the concept of over-potential, η, defined as:

$$\eta = E - E_e \tag{1.41}$$

The overpotential is the deviation in the applied potential from the equilibrium potential for the couple O/R. The use of overpotential has the effect of transposing the potential from an arbitrary scale determined by the choice of experimental reference electrode to a scale determined by the chemistry of the O/R couple itself. Remember that the current density at the equilibrium potential is zero and it can then be seen that the overpotential can be envisaged as the driving force for oxidation/reduction (depending on the sign of overpotential).

Substituting Equation (1.41) into (1.40) and then considering the situation at the equilibrium potential (putting $\eta = 0$) leads to:

$$j_0 = nFk_a^\circ c_R \exp\frac{\alpha_a nFE_e}{RT} = -nFk_c^\circ c_O \exp-\frac{\alpha_c nFE_e}{RT} \qquad (1.42)$$

since at this potential the partial anodic and cathodic current densities are both equal to the exchange current density. Now, using the equalities of (1.42) and Equation (1.41) in Equation (1.40) leads to the Butler–Volmer equation:

$$\begin{aligned}
j &= j_0\left(\exp\frac{\alpha_a nF\eta}{RT} - \exp-\frac{\alpha_c nF\eta}{RT}\right) \\
&= j_0\left(\exp\frac{\alpha_a nF\eta}{RT} - \exp-\frac{(1-\alpha_a)nF\eta}{RT}\right)
\end{aligned} \qquad (1.43)$$

which is central to our discussion of the kinetics of simple electron-transfer reactions.

It can now be seen that because of the opposite signs in the arguments of the exponentials, the net anodic current at a potential positive to the equilibrium potential will arise by an increase in the anodic partial current density and a decrease in the partial cathodic current compared to the situation at the equilibrium potential (Figure 1.3). Conversely, the

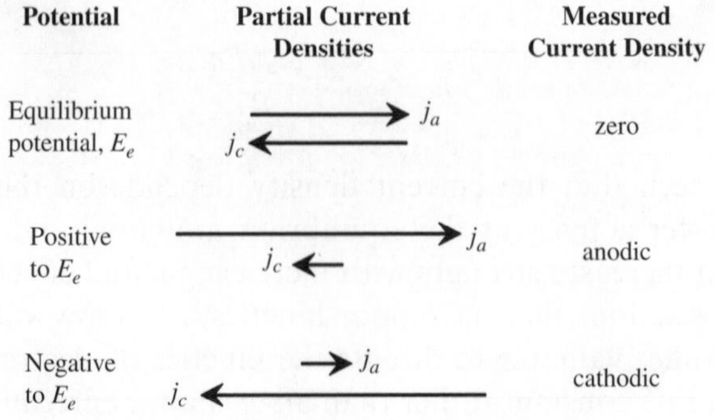

Potential	Partial Current Densities	Measured Current Density
Equilibrium potential, E_e	$j_c \leftarrow \quad \longrightarrow j_a$	zero
Positive to E_e	$j_c \leftarrow \quad \longrightarrow j_a$	anodic
Negative to E_e	$j_c \longleftarrow \quad \rightarrow j_a$	cathodic

Figure 1.3 Influence of the potential relative to the equilibrium potential (overpotential) on the partial anodic and cathodic partial current densities and the measured current density.

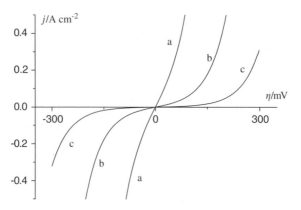

Figure 1.4   Influence of the exchange current density, j_o, on the j–η response when $n = 1$ and $\alpha_A = \alpha_c = 0.5$. $j_o =$ (a) 0.1, (b) 0.01 and (c) 0.001 A cm^{-2}.

net cathodic current at negative overpotentials arises by an increase in the cathodic partial current density and a decrease in the partial anodic current compared to the situation at the equilibrium potential. The general form of the current density *vs* overpotential response predicted by Equation (1.43) is illustrated in Figure 1.4; in fact, the responses for three values of the exchange current density are shown and it can readily be seen that the overpotential required to drive the electron-transfer reaction at a particular rate (current density) increases with decreasing exchange current density.

Equation (1.43) has two limiting forms commonly used in experimental electrochemistry. Firstly, only close to the equilibrium potential are both exponential terms in the equation significant. For a moderate positive overpotential (say, $\eta > 26$ mV), $j_a \gg j_c$ and the equation becomes:

$$j = j_a = j_o \exp \frac{\alpha_a n F \eta}{RT} \tag{1.44}$$

It can be seen that the current density depends on the amount of electron-transfer activity at the equilibrium potential and a term containing η that increases strongly with increasing η and effectively acts as a driver for reactions that have poor kinetics, *i.e.* a low value of j_o.

It is sometimes valuable to discuss the kinetics of electrode reactions in terms of a rate constant rather than an exchange current density. The standard rate constant, k_s, is the rate constant for both oxidation and reduction at the *formal potential* (in contrast, the exchange current density is the equal anodic and cathodic partial current densities at the *equilibrium potential*) and is given by:

$$j_0 = nFk_s(c_O)^{\alpha_a}(c_R)^{\alpha_c} \tag{1.45}$$

Equation (1.44) can also be written:

$$\log j = \log j_0 + \frac{\alpha_a nF}{2.3RT}\eta \tag{1.46}$$

and it is evident that for a reaction with $n = 1$ and $\alpha_a = 0.5$ the current will increase by a factor of ten for every increase in overpotential of 120 mV; overpotential is a very strong driver! In fact, for more complex reactions, the factor of ten increase may be obtained with an increase in overpotential of only 60, 40 or 30 mV. An overpotential of a volt can easily lead to an increase in rate between 10^8 and 10^{30} – this is massive compared to, for example, the influence of temperature on homogeneous chemical reactions and allows very slow reactions to be driven. A similar argument for negative potentials leads to:

$$\log -j = \log j_0 - \frac{\alpha_c nF}{2.3RT}\eta \tag{1.47}$$

Since a cathodic current will be observed at these overpotentials, $-j$ is a positive quantity. Equations (1.46) and (1.47) are known as Tafel equations; Figure 1.5 illustrates current density *vs* overpotential in this form. At both positive and negative overpotentials there is a range where the $\log |j|$ *vs* η plot is linear and both extrapolate to $\log j_0$ at $\eta = 0$.

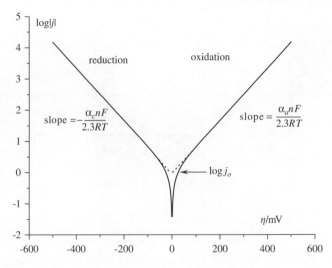

Figure 1.5 Tafel plots for the analysis of j *vs* η data to obtain the kinetic parameters for a simple electron-transfer couple O/R.

Moreover, the transfer coefficients may be obtained from the slopes and the slopes should be related since $\alpha_a + \alpha_c = 1$. For a reaction where $n = 1$ and $\alpha_a = \alpha_c = 0.5$, the Tafel slopes will be $1/120\,\text{mV}$.

The second limiting form of the Butler–Volmer equation, equation (1.43), applies at low vales of overpotential, *i.e.* very close to the equilibrium potential. Then, expanding the exponential terms as a series and neglecting squared and higher order terms, for $\alpha_a = \alpha_c = 0.5$ one obtains the very simple equation:

$$j = j_o \frac{nF}{RT}\eta \tag{1.48}$$

Very close to the equilibrium potential, there is a linear relationship between current and overpotential but the range of overpotential where the relationship is exact is limited. This is illustrated in Figure 1.6, where j/j_o is calculated from Equation (1.43) over a limited range of overpotential. Usually, the linear approximation is really precise only for $\eta < 10\,\text{mV}$.

When current density *vs* overpotential data are available for both oxidation and reduction, and over a wider range of overpotentials, the data can conveniently be presented on a single line using a rearranged form of Equation (1.43):

$$\log \frac{j}{\left(1 - \exp - \frac{nF\eta}{RT}\right)} = \log j_o + \frac{\alpha_a nF}{RT}\eta \tag{1.49}$$

Again, the exchange current density is found from the intercept at $\eta = 0$ and the transfer coefficient from the slope.

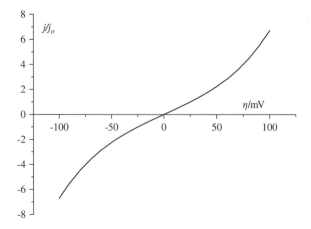

Figure 1.6 Plot of j/j_o calculated from Equation (1.43) *vs* η to show that j is a linear feature of overpotential only for a very limited range of overpotentials close to the equilibrium potential.

In summary, the rate of electron transfer or current density for a reaction is determined by:

- the exchange current density (or by the standard rate constant and concentrations),
- the transfer coefficient,
- the applied overpotential,
- temperature, largely through its influence on the exchange current density.

A major obstacle to understanding the kinetics of electron-transfer reactions is the terminology; as a further aid, Table 1.4 summarizes the important parameters used in this section.

1.4 MASS TRANSPORT

The introduction to this chapter emphasized that the supply of reactant and the removal of product from the electrode surface are essential to a continuing chemical change and current. In general, there can be contributions from three forms of mass transport.

1. *Diffusion* is the movement of a species due to a gradient in concentration. In other words, it is the physical process whereby

Table 1.4 Summary of the terminology used in the discussion of a simple electron-transfer reaction, $O + ne^- \rightleftarrows R$.

Symbol	Definition	Units
E	Experimental potential *vs* a reference electrode	V
E_e	Equilibrium potential for the couple O/R *vs* reference electrode	V
E_e°	Formal potential for the couple O/R *vs* reference electrode – the equilibrium potential when $c_O = c_R$	V
η	Overpotential $(= E - E_e)$	V
I	Experimental current	A
j	Experimental current density $(= I/A)$	A cm^{-2}
j_o	Exchange current density – the equal partial anodic and cathodic current densities at E_e	A cm^{-2}
j_a	Partial anodic or oxidation current density	A cm^{-2}
j_c	Partial cathodic or reduction current density	A cm^{-2}
α_a	Anodic transfer coefficient	–
α_c	Cathodic transfer coefficient	–
k_a	Rate constant for oxidation	cm s^{-1}
k_c	Rate constant for reduction	cm s^{-1}
k_s	Standard rate constant	cm s^{-1}
F	Faraday constant	96485 C mol^{-1}

nature minimizes differences in concentration, and diffusion will always occur from regions with high concentration of a species to regions that are more dilute until the concentration is uniform. Diffusion is inevitable with all electrode reactions since the electron transfer leads to a lowering of reactant concentration and an increase in product concentration at the surface compared to the bulk solution.

2. *Convection* is the movement of a species due to external mechanical forces. Convection may be introduced deliberately by shaking the cell, sparging the solution with gas, stirring the solution or moving the electrode. In many circumstances, however, it is desirable to be able to describe the convective regime quantitatively; this is possible only for systems with simple hydrodynamics and these include the rotating disc electrode and flow of the solution past a flat plate electrode. Normally, if the experiment involves forced convection, the rate of transport of species by convection totally dominates that by diffusion. The practical electrochemist must also be aware of 'natural convection' in unstirred solution. This arises from causes such as random vibrations in the laboratory or the small density gradient resulting from both concentration and temperature changes within the layer adjacent to the electrode associated with the electron-transfer reaction itself at the electrode surface.

3. *Migration* is the movement of charged species due to a potential field. In all electrochemical cells, current is driven through the solution between two electrodes and this requires the existence of a potential gradient. Migration is then the process whereby charge passes through the solution (see the discussion of Figure 1.1). Migration is purely an electrostatic phenomenon and is not necessarily an important mode of transport for either the reactant or product of the electrode reaction. Although, in all electron-transfer reactions, either the reactant or the product (or both) must be an ionic species, migration may not be an important form of transport for these species. In systems with a large excess of an inert electrolyte it will be largely the ions from this electrolyte that carry the charge through the solution. Indeed, this is another reason why most laboratory experiments are carried out with a high concentration of an electrolyte in solution. In contrast, in industrial electrolysis cells, the reactant may be charged and present in high concentration (*e.g.* a chlor-alkali cell, Figure 1.1) and then migration of the reactant must be taken into account.

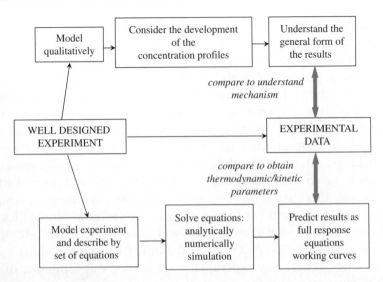

Figure 1.7 Procedures for the understanding and analysis of data from electro-chemical experiments.

Before proceeding further, it is important to recognize the principles underlying the interpretation of data from electrochemical experiments. Conclusions from practical electrochemistry are always based on a comparison of the experimental responses with predictions based on a model, as illustrated in Figure 1.7. These may be qualitative or quantitative but both are based on an understanding of the mass transport regime. Indeed, the latter requires that the regime may be described by a set of equations that may be solved; such experiments are therefore limited to a few mass transport regimes. The electrochemical experiment is introduced into the model *via* the surface concentrations and how they change with time; these are calculated from either the Nernst equation or the Butler–Volmer equation (Section 1.6). Fortunately, in general, the mathematics have been done earlier and the solutions are in the literature. The qualitative interpretation, however, greatly enhances understanding of our experiments and the use of equations from the literature requires us to be certain that the experi-mental mass transport regime is appropriate to the equations used. Again, in contrast, in industrial cells the objective is usually only to enhance the rate of mass transport, which can be achieved with a wide variety of approaches.

At this stage, it is useful to be very clear as to the objectives of laboratory experiments. As well as defining the overall chemical change, the aim must be to determine as much as possible about the mechanism and kinetics of the electrode process. The first stage will be to identify the key steps, including the rate-determining step; for a simple electrode reaction, this may be electron transfer or mass transport but it will also be necessary to develop tests to identify when adsorption, coupled chemical reactions and/or phase growth are involved and to identify features of the experimental responses that will disclose roles for such steps. Then, the next stage will be to determine parameters quantitatively, for at least the rate-determining step. In support of electrochemical technology, laboratory experiments usually have two clear goals, (a) to increase the selectivity of the reaction of interest when, for example, there are competing electron-transfer reactions or coupled chemical processes and (b) to modify the rate (current density) for the reaction of interest – usually the aim will be to increase the rate of the reaction of interest (the exception is the inhibition of corrosion).

In practice, laboratory experiments are carried out using one of two types of mass transport regime:

1. *Diffusion* only systems – The experiments are carried out with a still solution in a thermostat with a large excess of an inert electrolyte. Diffusion is the only form of mass transport that need be considered, at least until natural convection begins to interfere after some 10–100 s.
2. *Convection* as the predominant form of mass transport. In this book, such experiments will be illustrated by the rotating disc electrode. Migration as an influence on the transport of electro-active species will again be avoided by the addition of an inert electrolyte. Diffusion close to the electrode surface will still occur but commonly the introduction of convection increases the rate of transport by more than a factor of ten so that it is clearly the predominant mode of mass transport. In such experiments the regime is often called convective-diffusion.

1.4.1 Diffusion Only Conditions

The simplest model to describe such experiments is that of linear diffusion to a plane electrode. It assumes that the electrode is completely flat on a molecular scale and also of infinite dimensions so that concentration differences arise only in the direction perpendicular to the electrode surface. At first sight, this seems a very idealistic model since the electrode will not be completely flat and is certainly of finite size. Moreover, the common electrodes in the laboratory are discs, spheres, wires and spades. In fact, however, linear diffusion to a plane electrode is a very satisfactory model and several real electrode geometries may be shown by rigorous mathematics to be adequately approximated by it in experimental conditions.

The theoretical treatment of many electrochemical experiments is developed through calculating the way that concentrations of reactants and products change with distance from the electrode through a layer close to the surface that is disturbed by the electrode reaction – the so-called concentration profiles. Moreover, our understanding of our experiments is greatly enhanced by considering in a qualitative way how these concentration profiles change with time during experiments.

Diffusion is described quantitatively by Fick's laws (Figure 1.8). In the context of the model, linear diffusion to a plane electrode, these may be written in one dimension, namely, that perpendicular to the surface. Fick's first law discusses the rate of diffusion through a plane parallel to the electrode and at a distance, x, from the surface.

The rate of diffusion is known as the flux and it has the units mol $cm^{-2} s^{-1}$. Fick's first law states that the flux is proportional to the concentration gradient, dc/dx at the distance, x, from the electrode or:

$$\text{Flux} = -D\frac{dc}{dx} \tag{1.50}$$

where the proportionality constant is known as the diffusion coefficient, D ($cm^2 s^{-1}$). The minus sign ensures that the species diffuses from

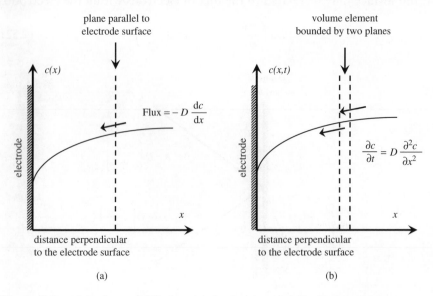

(a) (b)

Figure 1.8 Fick's laws of diffusion: (*a*) the first law and (*b*) the second law.

concentrated to dilute regions of the solution. In aqueous solutions, a typical value for a diffusion coefficient is $10^{-5}\,cm^2\,s^{-1}$. The second law considers the change with time of the concentration of the diffusing species at the centre of a volume element bounded by two planes parallel to the surface. The concentration changes result from diffusion into the element through one plane and out through the other. Fick's second law states:

$$\frac{\partial c}{\partial t} = D\frac{\partial^2 c}{\partial x^2} \tag{1.51}$$

The concentration is now a function of both distance and time, *i.e.* the concentration profiles, $c = f(x,t)$ will change with time as diffusion seeks to minimize differences in concentrations throughout space. The solution of this partial differential equation, together with initial and boundary conditions appropriate to the experiment, is the approach to developing a precise theoretical description of experiments. Generally, this is achieved through Laplace transform procedures although, in general, the solutions are readily available in textbooks.

An important application of Fick's law is to the situation at the electrode surface (Figure 1.9). The electrode reaction leads only to the interconversion of O and R and, since the law of conservation of matter must apply, the fluxes of O to the surface and R away from it must be equal. In addition, the conversion of one O into one R must be accompanied by the transfer of n electrons. Hence, the fluxes of O and R at the surface may be related to the flux of electrons within the electrode:

$$\frac{j}{nF} = D_R\left(\frac{\partial c_R}{\partial x}\right)_{x=0} = -D_O\left(\frac{\partial c_O}{\partial x}\right)_{x=0} \tag{1.52}$$

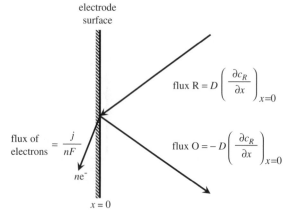

Figure 1.9 Balance of the fluxes at the electrode surface for the reaction $R - ne^- \rightarrow O$.

This shows the relationship between the current density and fluxes of O and R at the surface and leads to expressions for the current density from the solutions to Equation (1.51):

$$j = nFD_R \left(\frac{\partial c_R}{\partial x}\right)_{x=0} = -nFD_O \left(\frac{\partial c_O}{\partial x}\right)_{x=0} \tag{1.53}$$

In the steady state, diffusion profiles will always be linear; if the profiles are not linear, there will be some points in space where the concentration differences have not reached a minimum value and diffusion will continue until the concentration differences have been minimized everywhere.

Many experiments are carried out under conditions of non-steady state diffusion; with electronic instrumentation it is possible to change the electrode potential, and therefore the surface concentration of reactant, rather rapidly. In comparison, diffusion is a rather slow process and changes to the concentration profiles close to the electrode resulting from the change in surface concentration will occur over several seconds. The argument is best developed using a specific example:

> *Example* – In a cell (in a thermostat) with a still solution, containing the species R with concentration c_R and an excess of inert electrolyte, the potential of an electrode is stepped from a value at which the current density is zero (no chemical change is occurring) to one very positive to the equilibrium potential for the couple O/R.

From Equations (1.29) and (1.44) the change in potential is effectively an instruction to the electrode to change instantaneously the ratio of c_O/c_R at the surface to a very high value. This can happen only by the rapid conversion of R into O at the electrode (a high current density will be observed) and the concentration of R at the surface will drop from c_R to very close to zero). But this change in concentration is achieved immediately only at the electrode surface, $x = 0$. However, the change in concentration at the electrode surface has created concentration differences and diffusion will result, causing R to move towards the electrode. If the potential is held at the high positive overpotential, the surface concentration of R will remain close to zero and the concentration profile will develop with time. Diffusion will seek to minimize concentration differences at all distances from the surface. Inevitably, two trends will occur: (a) the flux of R to the surface will decrease with time and, since the observed current is proportional to this flux, the current density will drop significantly with time and (b) the thickness of the layer affected by the experiment (the diffusion layer) will increase as species away from the surface learn about the event at the electrode surface. Figure 1.10(a) shows the development of the concentration profiles during this experiment.

Such a qualitative discussion of the concentration profiles during the experiment is illuminating and leads to the conclusion that the response to the potential step is a transient where the current density decreases with time. However, to predict the exact form of the falling transient it is necessary to solve Fick's second law, Equation (1.51). Effectively, the mathematical procedure will have to carry out three integrations, one with respect to time and two with respect to distance. To evaluate the three integration constants it is necessary to specify the concentration of R at all distances at one time and two distances for all times. Fortunately, this information is available. At the instant the potential is stepped, $t = 0$, no chemistry has occurred and the concentration of R is uniform at the level prepared for the experiment, *i.e.*:

$$\text{at } t = 0 \text{ and at all } x, \; c_R(x, 0) = c_R \qquad (1.54)$$

In addition, following the potential step, the concentration of R at the electrode surface can be written as zero while a long way from the electrode, no chemical change has occurred, *i.e.*:

$$\text{for } t > 0, \quad \text{at } x = 0, \quad c_R(0, t) = 0 \quad \text{and at } x = \infty, \; c_R(\infty, t) = c_R \qquad (1.55)$$

The solution of this set of equations (found by Laplace transformation procedures) leads to:

$$j = \frac{nFD^{1/2}c_R}{\pi^{1/2}t^{1/2}} \qquad (1.56)$$

This is known as the Cottrell equation and the theoretical transient is shown in Fig 1.10(b). In the experiment discussed, all information about electron transfer has been lost by the choice of potential and the only parameter that may be determined relates to diffusion, the diffusion coefficient. The discussion leads to (a) the exact form of the falling j *vs* t transient, Equation (1.56), (b) tests to determine whether an electrode reaction is diffusion controlled at an experimental potential and (c) methods to determine the diffusion coefficient; (b) and (c) can be achieved in several ways:

- Plotting j *vs* $t^{1/2}$ and obtaining a straight line passing through the origin confirms that the reaction is diffusion controlled – D is obtained from the slope.

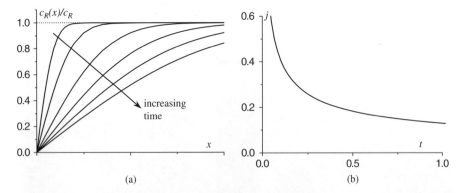

(a)   (b)

Figure 1.10   (a) Development of concentration profiles following a potential step from a value where $j = 0$ to one where the oxidation of R to O is diffusion controlled; (b) the corresponding current density *vs* time transient.

- Demonstrating that $jt^{1/2}$ is a constant also shows that the reaction is diffusion controlled – D is then calculated from this value.
- Plotting Equation (1.56) for various values of D and fitting to the experimental data.

To study the kinetics of electron transfer, it would be necessary to step to a lower overpotential where the surface concentration is not zero. Fick's second law would then have different boundary conditions at $x = 0$ and the j *vs* t transient would have a different shape (Chapter 7).

Another clear conclusion from this discussion is that the rate of non-steady state diffusion is a function of time; by carrying out experiments at short time, the rate of diffusion to the electrode can be increased substantially. This concept is central to the way electrochemical experiments are carried out. Exploitation of time in chronoamperometry, scan rate in cyclic voltammetry and frequency in ac impedance allows experiments to be carried out with different rates of diffusion and this is central to investigating the kinetics of both electron transfer and coupled chemical reactions. This will become clearer in Chapter 7.

1.4.2   Convective-diffusion Conditions

Under these conditions, the dominant form of mass transport will be convection; in the laboratory it is quite possible to design experiments where the introduction of convection increases the rate of mass transport by more than a factor of a hundred. For mechanistic and kinetic studies, it is important that the hydrodynamics of the system can be described precisely. The example used here is the rotating disc electrode.

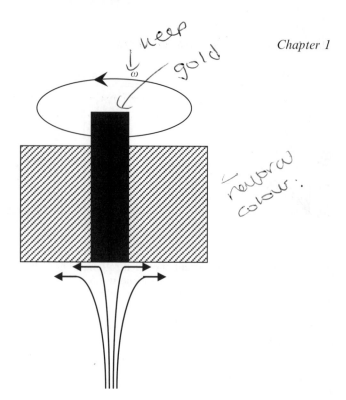

keep
gold
ω

neubral
colour :

Figure 1.11   Solution flow induced by the rotation of a disc electrode (with insulating sheath) in the electrolyte.

The rotating disc electrode consists of a polished disc of the selected electrode material surrounded by an insulating sheath of significantly larger diameter. Figure 1.11 illustrates the solution movement introduced by the rotation of the disc electrode. The principal movement is towards the disc, perpendicular to the surface. But since the solution cannot pass through the solid electrode/sheath surface, close to the surface the solution is thrown outwards and then circulates back into the bulk volume. As far as the disc is concerned, the mass transport approximates well to a one-dimensional regime where the critical direction is perpendicular to the surface. It can also be seen that the rate of transport of species to the electrode surface will be determined by the rotation rate of the disc, ω; the higher the rotation rate, the stronger the stirring of the solution and the higher the rate of transport. Chapter 7 covers applications of the rotating disc electrode; here we shall only outline a widely used model for the mass transport to the rotating disc electrode, namely, the Nernst diffusion layer model.

Figure 1.12 sets out the model, where it can be seen that the solution is divided into two regions: (a) the bulk solution that is strongly mixed by the stirring and thereby maintains a constant concentration of the

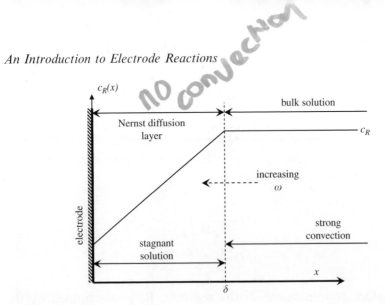

Figure 1.12   Nernst diffusion layer model for the oxidation of R to O at a rotating disc electrode. The bulk solution contains no O.

reactant throughout this region and (b) within a boundary layer adjacent to the surface, thickness δ, the solution is totally stagnant and diffusion becomes the only mode of mass transport. Clearly, this is an oversimplification. The hydrodynamics cannot change from a regime of strong stirring to a still solution immediately at $x = \delta$. However, the solution away from the disc is strongly and uniformly pumped towards the surface and the solution decelerates as it approaches the surface until the flow rate in the x-direction is zero at the solid surface. Hence, the Nernst diffusion layer model may be regarded as a useful equivalence model although it must be recognized that the layer thickness, δ, has no physical reality. Fortunately, the conclusions from the model are supported by a more precise mathematical model. The Nernst diffusion layer model, however, allows us to understand the key features of voltammetry at a rotating disc electrode.

Firstly, the response will depend on the rotation rate of the disc. In the model, stronger stirring of the bulk solution equates with a decrease in the thickness of the boundary layer. The precise mathematical model concludes that the thickness of this 'equivalent' boundary layer is given by:

$$\delta = \frac{1.61 \nu^{1/6} D^{1/3}}{\omega^{1/2}} \tag{1.57}$$

where ν is the kinematic viscosity (*i.e.* viscosity/density) of the solution. It can be seen from Figure 1.12 that increasing the rotation rate will increase the flux of reactant to the surface and hence the current density. Notably, the thickness of this mass transport layer is of the order of microns; the double layer discussed in Chapter 3 is quite different and it is a layer with a thickness of a few molecular dimensions, *i.e.* nanometres.

Secondly, it is now possible to write simple expressions for the current density. In the steady state, the concentration profiles within the stagnant layer must again be linear and hence the current density (for the oxidation of R to O) is given by:

$$
\begin{aligned}
j &= nFD_R \left(\frac{dc_R}{dx} \right)_{x=0} \\
&= nFD \frac{c_R - (c_R)_{x=0}}{\delta}
\end{aligned}
\tag{1.58}
$$

The surface concentration of the reactant, R, is determined by the potential applied and, at a high enough overpotential, this effectively becomes zero. Increasing the overpotential further cannot further change the surface concentration and the current density will become independent of potential. This limiting current density, j_L, represents the highest rate at which R can be converted into O. The limiting current density is given by:

$$
j_L = \frac{nFDc_R}{\delta}
\tag{1.59}
$$

Thirdly, it can be seen why a voltammogram at a rotating disc electrode is sigmoidal. Figure 1.13(a) shows the way in which the concentration profiles change with applied potential. Only the surface concentration changes with potential; at the equilibrium potential, the

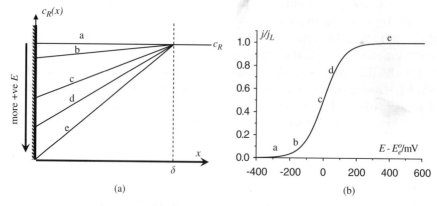

Figure 1.13 (*a*) Changes to the steady state concentration profiles at a rotating disc electrode during a voltammogram for the oxidation of R to O (only R in the initial solution). (*b*) The corresponding voltammogram. Labels a–e relate the concentration profiles to the potentials on the voltammogram.

concentration at the surface will be the same as the bulk, while as the potential is made more positive the rate of conversion of R into O will increase, thereby reducing the surface concentration of R towards zero when no further change can occur. While the surface concentration is decreasing the current will increase, but once it reaches zero no further decrease is possible and the current will plateau (Figure 1.13b).

In both Equations (1.58) and (1.59) it is possible to use Equation (1.57) to replace δ by measurable quantities. For example, with Equation (1.59), one obtains an expression known as the Levich equation:

$$j_L = \frac{1.61 n F D_R^{2/3} c_R \omega^{1/2}}{\nu^{1/6}} \tag{1.60}$$

It shows that a mass transport controlled current density is proportional to the square root of the rotation rate of the disc.

Notably, under all mass transport regimes, a universal expression for the mass transport controlled current density is:

$$j_L = n F k_m c \tag{1.61}$$

where k_m is a rate constant describing the rate of mass transfer. It is known as the mass transfer coefficient and can be determined experimentally from measurements with a known mass transport controlled reaction. At a rotating disc electrode, comparison of Equations (1.60) and (1.61) shows that:

$$k_m = \frac{1.61 D^{2/3} \omega^{1/2}}{\nu^{1/6}} \tag{1.62}$$

1.5 INTERACTION OF ELECTRON TRANSFER AND MASS TRANSPORT

So far we have largely treated electron transfer and mass transfer as independent processes but at the beginning we noted that the overall rate of the sequence:

$$R_{soln} \xrightarrow{\text{mass transfer}} R_{electrode} \xrightarrow{\text{electron transfer}} O_{electrode} \xrightarrow{\text{mass transfer}} O_{soln} \tag{1.63}$$

will be determined by the slowest step. Hence, understanding a complete current density *vs* potential response requires the recognition of how the steps interact. Again, we shall consider a particular experiment:

The experiment will be a steady state potential scan with an inert rotating disc electrode in a solution containing both O and R, with

$c_O/c_R = 0.1$, and an excess of inert electrolyte. We shall also assume that the electron-transfer reaction has poor kinetics so that the surface concentrations are controlled by the Butler–Volmer equation.

For the oxidation of R to O, four situations may be recognized:

1. *At the equilibrium potential* the current density will be zero – no net chemical change will occur. Because of the ratio of concentrations in the solution, the equilibrium potential will be negative to the formal potential, in fact it will be $[E_e^\ominus - (60/n)]$mV [Equation (1.29)].
2. *Close to the equilibrium potential* the rate of electron transfer will be very slow and clearly the rate-determining step. Equations (1.43–1.48) will be obeyed; a conclusion to the discussion of electron transfer was that the rate depends very strongly on potential. For $\eta > 50$ mV, the current density will increase exponentially with overpotential (and commonly by a factor of ten for each 120 mV) and a plot of $\log j$ vs η will be linear. In this potential range, the surface concentration of R will not deviate significantly from the bulk value. To understand this statement, it should be remembered that we can measure current over many orders of magnitude and when $j = 0.01 j_L$ the surface concentration will differ from that in the bulk by only 1%. The overpotential range where the simple linear $\log j$ vs η plot will be seen will depend on the kinetics of the electrode transfer reaction and the mass transport regime; in practice, it should be observed for the range $5 j_o < j < 0.05 j_L$.

   The solely electron-transfer controlled regime may be recognized experimentally; the current density:
   (i) varies strongly with potential;
   (ii) is independent of the mass transport regime (tested by varying the rotation rate of the disc, bubbling gas or even shaking the cell).
3. *At very high overpotential* the rate of electron transfer must increase to a very high rate and mass transfer will become the rate-determining step. The surface concentration of R will drop to zero and Equations (1.59–1.61) will apply. Pure electron transfer may also be recognized experimentally; the current density:
   (i) is independent of potential;
   (ii) varies strongly with the mass transport regime.
4. *Intermediate overpotentials*, where the rates of electron transfer and mass transfer are similar. The simple equations will not be obeyed and this situation corresponds to the case where the surface

concentration is significantly different from c_R but has not yet reached zero. This range of overpotentials extends from where electron transfer is clearly much slower than mass transfer to where the reverse is true. This requires a change in overall rate, say of 25 if a ratio of rates of 5 is considered enough to give a single rate-determining step. Hence one might expect to see mixed control over the range $j_L/25 < j < j_L$. It can be seen that the whole of the steeply rising portion of the voltammogram corresponds to mixed control; obtaining information about electron transfer or mass transport from this region will always be more complex than from potential regions with a single rate-determining step, *i.e.* the very foot of the wave and the plateau region. Experimentally, the current density in the region of mixed control:

(i) Varies with potential but less strongly than for pure electron-transfer control. With increasing overpotential, there will be a gradual change from an almost exponential dependence on overpotential to no variation.

(ii) Varies with the mass transport regime but less strongly than for full mass transfer control. Again, with increase in over-potential, the variation will be gradual from no dependence on the mass transport regime to that for full mass transfer control.

Figure 1.14 shows the full current *vs* potential characteristic for the solution where $c_R = 10c_O$ both as a plot of j *vs* E and $\log j$ *vs* η. We have discussed above only oxidation but the same arguments will apply to the reduction portion of the responses. Because of the differences in reactant concentrations, the rate of mass transfer of R and O will differ by a factor of 10 and, in consequence, the mass transfer limiting current densities will also differ by a factor of 10. The regions of electron-transfer control, mixed control and mass transport control are marked on the two curves. The separation of the two waves in the j *vs* E response is a function of the kinetics of electron transfer; the waves separate with decreasing values of j_o (or k_s) since the overpotential to drive both oxidation and reduction will increase.

1.6 REVERSIBLE *VS* IRREVERSIBLE ELECTRODE REACTIONS

In electrochemistry, the terms 'reversible' and 'irreversible' are used in the tradition of thermodynamics. They are used to discuss whether

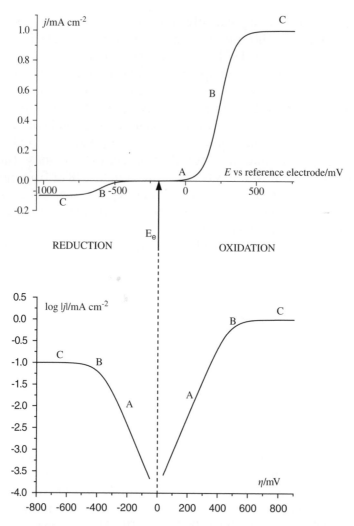

Figure 1.14 Current *vs* potential data for a simple electron-transfer reaction, $O + ne^-$ → R, presented as both *j vs E* and log *j vs η*. A: electron-transfer control, B: mixed control and C: mass transfer control.

the electron-transfer reaction at the surface is in thermodynamic equilibrium.

So far, we have discussed only the case where the electron transfer is slow compared to mass transfer. Then the electron-transfer reaction is not in equilibrium; the surface concentrations of O and R are determined by kinetic equations and an overpotential is required to drive the reaction at a particular rate. Electrode reactions where this treatment is appropriate are termed 'irreversible'. There is, however, another

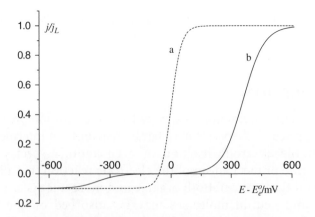

Figure 1.15 Plot of normalized current density *vs* potential response at a rotating disc electrode for (a) a reversible electrode reaction and (b) an irreversible electrode reaction. Solution contains $c_R = 10c_O$.

possibility. If the electron-transfer step is inherently fast (high k_s and j_o), then it is possible that, at all potentials and under the prevailing mass transfer regime, the electron-transfer reaction on the surface remains in equilibrium. Then, the surface concentrations may be calculated from the Nernst equation, which is purely a thermodynamic equation; no overpotential is necessary to obtain any rate of conversion between O and R. Such electrode reactions are termed 'reversible'.

Figure 1.15 shows the voltammograms obtained at a rotating disc electrode for both types of electrode reaction. As noted previously in the discussion of Figure 1.14, the voltammogram for an irreversible couple has two well-separated waves, one for oxidation and one for reduction, with the separation depending on the kinetics of the couple O/R. In contrast, the voltammogram for a reversible couple shows a single wave, the response crossing the zero current axis directly between oxidation and reduction because no overpotential is necessary to drive the electron transfer. The wave for a reversible couple is also much steeper.

Notably, however, whether a reaction appears reversible or irreversible will depend on the mass transport regime. For reactions with intermediate kinetics, it is possible that the electrode reaction appears reversible with experimental conditions where the mass transport regime is relatively poor but a substantial increase in the rate of mass transport (*e.g.* by increasing the rotation rate of a disc electrode or working on a shorter timescale) may lead to the reaction becoming irreversible. This is the concept underlying methods to study fast electron-transfer reactions (Chapter 7). More quantitatively, the distinction between reversible and irreversible depends on the relative values of k_m and k_s, the mass transfer coefficient for the experimental conditions and the standard rate constant for the couple O/R, respectively.

1.7 ADSORPTION

Adsorption is the interaction of species from the solution phase with the electrode surface. The species may be the reactant, an intermediate or a product of the electrode reaction or even another species added to change the rate or mechanism of the electrode reaction (usually by itself adsorbing on the surface). Both organic and inorganic species as well as both ions and neutral molecules may be adsorbed on the electrode. Certainly, the solvent and ions of the inert electrolyte will adsorb under many conditions.

The nature and strength of the interaction vary widely. The strongest arises from the formation of a covalent bond, *e.g.*:

$$H^+ + Pt_{site} + e^- \longrightarrow Pt_{site} - H \qquad (1.64)$$

$$CH_3OH + 2Pt_{sites} \longrightarrow Pt_{site} - CH_2OH + Pt_{site} - H \qquad (1.65)$$

but there is also a range of interactions resulting from electrostatic forces. At each applied potential, the electrode surface will have a characteristic surface charge that is also dependent on the electrode material and the solution composition. This surface charge leads to the attraction of ions, dipoles and π-electron systems. Moreover, dipoles may be induced by the electric field at the electrode/solution interface. All adsorption processes may be reversible or irreversible and Reactions (1.64) and (1.65) are examples of these two possibilities.

Species may be adsorbed from solution onto a surface at open circuit but the richness of the effects observed is enhanced by the variation of charge with applied potential; the surface charge may readily be made positive or negative and the charge density varied from low to high. The potential where the surface charge changes from positive to negative is known as the potential of zero charge.

The extent of adsorption, often expressed as a fraction of the surface covered by the adsorbate, θ, is best understood in terms of two competitions: (a) between the surface and the solution for a potential adsorbate – thus, for example, organics are less likely to adsorb from an organic medium than water; (b) between all species in the solution (reactant, solvent, ions from the electrolyte, additives and impurities) for the finite number of sites on the electrode surface. The tendency to adsorb is expressed as a Gibbs free energy of adsorption but this depends on both electrode material and the solution medium. The coverage by adsorbate also depends on the potential of the electrode; ions are most likely to absorb on a surface of opposite charge while neutral molecules tend to adsorb most strongly at the potential of zero charge where there is no competition from ions.

1.7.1 Study of Adsorption

Many adsorption processes are reversible (*i.e.* rapid and therefore in equilibrium) and then the coverage is conveniently discussed in terms of isotherms. These relate the coverage, θ, at constant temperature, to the concentration of adsorbate in solution and the Gibbs free energy of adsorption. Many isotherms have been proposed and they differ in the extent and method of taking into account the lateral interaction between adjacent adsorbate species on the surface. The simplest isotherm is the Langmuir isotherm and this assumes that there are no lateral interactions (*i.e.* the Gibbs free energy of adsorption is independent of coverage). It is written:

$$\frac{\theta}{1-\theta} = c \exp\frac{-\Delta G^{\circ}_{ads}}{RT} \tag{1.66}$$

On the other hand, the Frumkin isotherm is based on the assumption that it becomes more difficult to adsorb further species as the coverage increases – in fact the Gibbs free energy increases linearly with coverage:

$$\Delta G_{ads} = \Delta G^{\circ}_{ads} + g\theta \tag{1.67}$$

$$\frac{\theta}{1-\theta} = c \exp\frac{-\Delta G^{\circ}_{ads}}{RT} \exp\frac{-g\theta}{RT} \tag{1.68}$$

When the adsorbate is electroactive or results from an electron-transfer reaction, the coverage can be assessed by recording a cyclic voltammogram and determining the charge under the peak for the electrode reaction. The peak shape for the oxidation/reduction of an adsorbed species is different from that for a solution-free species and the potential for the reaction will also be shifted compared to that for the same reaction with the reactant/product in solution (Chapter 7). For example, adsorption of the reactant on the electrode surface will stabilize the species to electron transfer and will make oxidation/reduction more difficult; oxidation of the adsorbed species will take place positive to the solution species and reduction negative to the solution species. Adsorption of the product will allow the reaction to occur more easily and the opposite shifts are observed. If the adsorbate is not electroactive, at least in the potential range of interest, the surface coverage is normally deduced using AC techniques to determine the capacitance as a function of the concentration of the species in solution and potential (Chapter 3).

When the adsorption process is irreversible, the coverage cannot be discussed in terms of an isotherm. But such reactions are important in fuel cells. Equation (1.65) is the first stage in the oxidation of methanol at a fuel cell anode. It is followed by a series of further steps that involve other adsorbed organic fragments as well as adsorbed carbon monoxide. Each of the adsorbates may be oxidized to CO_2 or decompose in competing pathways so that the coverage by each species depends on the kinetics of several steps. *In situ* spectroscopic techniques, particularly infrared and mass spectrometry directly coupled to the electrochemistry, have aided our understanding of such reactions but, unsurprisingly, such complex systems cannot yet be fully defined. In simple situations where only one adsorbed species is present, coverage can usually be determined from the charge for oxidation/reduction.

1.7.2   Why is Adsorption of Interest?

1.7.2.1   Electrocatalysis.   Catalysis is the enhancement of the rate of a reaction by a species that is not consumed in the overall reaction sequence; the role of the catalyst is to provide an alternative, low energy of activation pathway for the conversion of reactant into product. In electrocatalysis, the catalyst is usually the electrode material itself and the mechanism by which it increases the rate usually involves adsorption of reactant or intermediates. Experimentally, the role of the electrocatalyst is to (a) increase the current density at a fixed potential or (b) to reduce the overpotential to support the reaction at a fixed current density. Because of the importance of fuel cells, important reactions requiring electro-catalysis include oxygen reduction and hydrogen or methanol oxidation. These systems are discussed further in Chapters 5 and 8.

1.7.2.2   Inhibition of Electron Transfer.   The adsorption of molecules not directly involved in the electrode reaction can inhibit the electron-transfer reaction. The best known example is corrosion inhibitors. The role of the inhibitor can be modelled in two ways. Firstly, it can be envisaged that the adsorbate completely covers the surface or, at least, all the active sites, thereby increasing the distance over which the electron must hop between the electrode and reactant. This is a possible mode of operation of inhibitors that slow down iron dissolution, *i.e.* anodic inhibitors. Secondly, the mode of action may involve competing for sites on the surface with an intermediate in a reaction; for example, cathodic corrosion inhibitors compete with adsorbed hydrogen atoms for sites on a steel surface.

1.7.2.3 Additives in Electroplating. In the electrodeposition of metals, additives are widely used to control the form of the deposit. This may, for example, be its smoothness or brightness, the morphology of the deposit, or the size or shapes of the crystallites. The additives are thought to act by adsorbing on particular sites on the surface.

1.7.2.4 New Chemistry. Additives can direct an electrode reaction down different pathways leading to different products. An example is an industrial process for the hydrodimerization of acrylonitrile to adiponitrile, Reaction (1.11) in Table 1.2. This reaction requires the presence in the medium of tetraalkylammonium ions. In their absence, hydrogen evolution is a major competing electrode reaction and the main organic product is propionitrile. The tetraalkylammonium ions are thought to adsorb on the cathode surface and create a local environment with a low proton donating ability.

1.8 COUPLED CHEMICAL REACTIONS

It is very common for the product of electron transfer to be unstable and for it to undergo a chemical reaction, *e.g.*:

$$O + ne^- \longrightarrow R \tag{1.69}$$

$$R \xrightarrow{k} P \tag{1.70}$$

where for the moment it will be assumed that P is electroinactive and stable. The chemical step may occur either as a heterogeneous reaction while R is adsorbed on the electrode surface or as a homogeneous reaction in the electrolyte medium while R is moving away from the electrode surface. Only the case of a homogeneous reaction will be considered in this section. In addition, the primary electrode reaction is written as a reduction. Naturally, a similar section could be written with oxidation as the primary step.

Importantly, the chemistry of R will generally be unchanged by its origin at the electrode surface; when present at a similar concentration, the reaction pathways and products will be the same as when it is formed by a chemical reaction in the same medium as used for the electrolysis (although, unlike solutions for chemical reaction, electrolysis media usually contain high levels of electrolyte and this may effect the chemistry of charged intermediates). Commonly, the mechanism and kinetics of the reactions of R are the same, and then electroanalytical techniques provide a way to study the homogeneous chemistry of R. Of course, as always with reactive intermediates, there is no certainty that the chemistry of R will lead to a single product and its decay can often involve competitive pathways. In addition, the rate constant for Reaction (1.70) determines the thickness of a reaction layer at the electrode surface. If the rate constant is high, all the chemistry occurs within a thin layer close to the electrode surface and the chemical reaction will influence experimental data strongly. In contrast, if the rate constant is low, the reaction layer is thick and the intermediate may even survive into the bulk solution; the chemistry then has less influence on experimental data for the electrochemical experiment.

In many systems, the product of the chemical reaction, P, may itself be electroactive and may undergo further reduction at the potential where it is formed or at more negative potentials or, indeed, it may be oxidized (back to O or an entirely different species) at a more positive potential. Such complexities add much to the richness of experimental electrochemistry. For example, cyclic voltammograms are observed with multiple peaks and the dependence of the additional peaks on potential scan rate and potential range provides much information about the chemistry. Another important type of reaction is:

$$O + ne^- \longrightarrow R \tag{1.71}$$

$$R + Q \xrightarrow{k} O + P \tag{1.72}$$

where the product of electron transfer reduces (or oxidizes) an electroinactive molecule with regeneration of the reactant for the electrode reaction; O can be considered as a catalyst for the conversion of Q into P. In the laboratory, these are known as 'mediated reactions' while in electrochemical technology they are more usually called 'indirect electrode reactions'. The mediator couple O/R may be inorganic (*e.g.* Ce^{4+}/Ce^{3+}, $Cr_2O_7^{2-}/Cr^{3+}$, Br_2/Br^-, Zn^{2+}/Zn), organic or an enzyme.

In view of the large variety of mechanisms observed, a nomenclature has been developed to describe mechanisms concisely (Table 1.5).

Table 1.5   Notation used to described complex reaction mechanisms, usually where the chemical steps are homogeneous in the electrolysis medium; the letter 'e' indicates an electron-transfer step and 'c' a homogeneous chemical step.

---

ec mechanism
$O + e^- \to R$
$R \to P$
ec' mechanism
$O + e^- \to R$
$R + Q \to O + P$
ece mechanism
$O + e^- \to R$
$R \to P$
$P + e^- \to Q$
ecec mechanism
$O + e^- \to R$
$R \to P$
$P + e^- \to Q$
$Q \to T$
ce mechanism
$M \to O$
$O + e^- \to R$

Throughout, 'e' stands for an electron-transfer reaction and 'c' for a homogeneous chemical step.

Electrolysis has been used for the synthesis of several chemicals, both organic and inorganic – both in the laboratory and on an industrial scale. Commonly, the isolated products result from a complex sequence of chemical steps and also multiple electron transfers. In particular, the electrode reactions of organic molecules almost always involve $2e^-$ steps since the cleavage or formation of bonds involves two electrons. Hence, to form single products in high yield it is necessary to control (a) the electrode potential so that the electrode reaction forms a single intermediate (either R or P could undergo further reduction if the potential is uncontrolled) and with a controlled flux from the electrode surface; this flux (and the current density) will also depend on the reactant concentration and the mass transport regime; and (b) the solution conditions (solvent, electrolytes, pH, the presence of other reactants, temperature, *etc.*) so that the chemistry of R in solution is selective.

Usually, the conditions for electrolysis are determined by a mixture of experience and a parametric study. Variables to be considered will include:

- electrode potential (although perhaps controlled indirectly through use of a controlled current and knowledge of the relationship

between j and E for the particular reactant concentration and mass transport regime);
- electrode material;
- solvent, electrolyte and pH;
- the mass transport regime;
- electroinactive reactants to trap intermediates formed at the electrode;
- additives (adsorbates) to modify the environment at the surface;
- the cell design – divided or undivided, geometry and form of the electrode, mass transport regime.

In practice, any experimental variable may influence several steps in the sequence. For example, temperature will influence the rate of mass transport [D will increase 1–2% per degree Kelvin (K)], the rate of electron-transfer reactions and the rate of all the chemical steps. Since the influence of temperature on the rates of competing reactions will not be the same, its influence on selectivity has to be determined experimentally. In the development of a commercial process, the selection of conditions will also depend on the relative priorities placed upon performance factors such as (a) the rate of conversion (current density), determining the initial cost of purchasing cells and (b) the reaction selectivity, the extraction procedure and energy consumption, determining the running costs. It should also be recognized that in an electrolysis cell the objective is rapid chemical conversion of reactant into product. Hence, the reactant will be present in high concentration and the cell will be designed with a large ratio of electrode area to electrolyte volume and to give a high rate of mass transport. This is quite different from the cell used for experiments to investigate the mechanism and kinetics (Chapter 6).

Returning to the discussion of an ec system, for example Reactions (1.69) and (1.70), it is necessary to consider how the chemical step is introduced into the theoretical description of an experiment. Figure 1.16(a) shows the concentration profiles for both O and R at one instant during an experiment where the reduction of O to R is diffusion controlled, R is stable and the initial solution contains O but not R. The concentration profile for the reactant O was discussed earlier. For the product R, its concentration will be high at the surface where it is being formed and zero in the bulk solution. In fact, the exact form of the concentration profile for R is readily deduced; since both O and R are stable and the only chemistry is the interconversion of O and R at the electrode, the sum of the concentration of O and R must be equal to the initial concentration of O at all distances from the electrode surface. If, however, R is unstable and undergoes a first-order, homogeneous

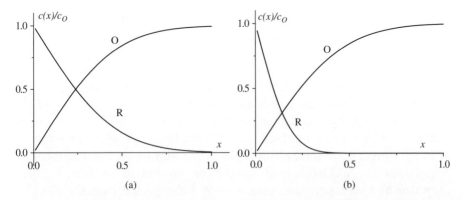

Figure 1.16 Instantaneous concentration profiles for O and R during an experiment where the potential is stepped from one where no reaction occurs to one where the reduction of O is diffusion controlled; (a) the intermediate, R, is stable and (b) R undergoes a first-order chemical reaction.

chemical reaction, the concentration of R will be lower at all x because the reaction is taking place with a rate kc_R throughout the reaction layer (Figure 1.16b).

Quantitatively, in a still solution, the change in concentration at any point in the reaction layer will depend on two factors: (a) diffusion into and out of the segment centred on x, *i.e.* Fick's second law, and (b) the rate of the chemical reaction. Hence:

$$\frac{\partial c_R(x, t)}{\partial t} = D \frac{\partial^2 c_R(x, t)}{\partial x^2} - kc_R(x, t) \tag{1.73}$$

The electroactive species O is not involved in the homogeneous chemistry and therefore changes in concentration arise only because of diffusion. Hence:

$$\frac{\partial c_O(x, t)}{\partial t} = D \frac{\partial^2 c_O(x, t)}{\partial x^2} \tag{1.74}$$

Although it might not be necessary for understanding the electro-chemical response, it is also possible to write an equivalent equation for the final product, P:

$$\frac{\partial c_P(x, t)}{\partial t} = D \frac{\partial^2 c_P(x, t)}{\partial x^2} + kc_R(x, t) \tag{1.75}$$

To complete the description of any experiment, one initial and two boundary conditions for each species will be required to allow integration of Equations (1.73) and (1.74). These will depend on both the solution prepared for the experiment and the nature of the electrochemical experiment.

In practice, the importance of the coupled chemical reaction in determining the response to the electrochemical experiment will depend on the relative values of the half-life of the intermediate R ($\tau_{1/2} = (\log 2)/k$) and the timescale of the experiment (*e.g.* in cyclic voltammetry this is determined by the potential scan rate). If the timescale of the experiment is much longer than the half-life of R, then almost all the intermediate formed at the electrode will have been consumed in the chemical step during the experiment and it is to be expected that the electrochemical response will be strongly changed by the coupled chemistry. In contrast, if the electrochemical experiment is completed in much less time than the half-life, the response will know little of the coupled chemistry. Thus, the concept underlying electrochemical methods to study coupled chemistry is to monitor the response over a range of experimental timescales.

1.9 PHASE FORMATION AND GROWTH

Electrode reactions that lead to the formation of a new solid phase on the electrode surface are surprisingly common. Examples include:

1. cathodic deposition of metals, *e.g.* Cu, Ni, Cr;
2. anodic deposition of conducting oxides, *e.g.* PbO_2;
3. deposition of conducting polymers, polypyrrole, polyaniline;
4. anodic formation of metal salts, *e.g.* AgCl, $HgSO_4$;
5. anodic formation of passivating oxides, TiO_2, Al_2O_3;
6. deposition of insulating organic films by electropolymerization or electropainting.

The type and extent of conductivity is the major factor in determining the electrochemical response from any experiment but the formation of a solid phase always significantly influences the electrochemical response. For example, the formation of an insulating film will always quickly lead to a very low current density and such processes are usually very irreversible.

A more interesting phase formation process is illustrated by the deposition of copper onto a carbon surface from a sulfuric acid bath:

$$Cu^{2+} + 2e^- \longrightarrow Cu \qquad (1.76)$$

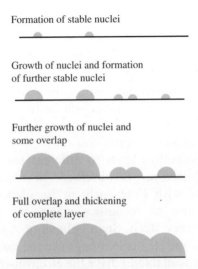

Formation of stable nuclei

Growth of nuclei and formation
of further stable nuclei

Further growth of nuclei and
some overlap

Full overlap and thickening
of complete layer

Figure 1.17   Electrodeposition of a metal layer by a mechanism involving progressive nucleation and growth as hemispherical centres.

It involves several distinct stages:

- nucleation of the new phase;
- growth of the individual metal centres;
- overlap of the growing centres, leading to a complete layer;
- thickening of the complete layer.

This is illustrated in Figure 1.17 for the particular case of continuous nucleation of hemispherical centres. Nucleation is the formation of stable centres. Most chemists will know from their attempts to crystallize or recrystallize a salt or compound that the formation of nuclei of a new phase is always an improbable event that must be forced to occur, *e.g.* by seeding or supersaturating the solution. This is because small centres are always unstable and tend to redissolve; the surface is unstable while stability comes from the interactions of atoms/ions/molecules in the bulk of the material. In the electrodeposition of metal, the nuclei consist of the metal atoms and when the centres are small the ratio of surface area to volume tends to lead to their redissolution. Formation of nuclei large enough to be stable is driven by the application of an overpotential in addition to that required to drive the electron-transfer reaction. There are also two limiting cases of the kinetics of nucleation: (a) instantaneous nucleation, in which nucleation occurs at all available sites, and (b) progressive nucleation, where the number of centres increases with time, following first order kinetics. The number density of nuclei can also vary strongly and it will depend on the electrode material and its pretreatment and the solution conditions as well as the metal being deposited. Once formed as stable entities, the individual centres grow with a characteristic shape such as hemispherical or conical until they are large enough to overlap with other expanding centres on the surface. Eventually, a complete layer will be formed and this will then thicken. At all of these stages, the reduction of cupric ion may be under electron transfer or mass transport control.

Each of these possibilities leads to different responses. In all cases, however, distinct behaviour is observed; at constant potential the current will increase with time early in the deposition process and voltammetry gives rise to responses where the current density *vs* potential curves are very steep. Such behaviour results from the over-potential required to initiate nucleation; once the nuclei are formed, this overpotential is then available to drive the electron-transfer reaction at a higher rate. In addition, as each centre grows, the surface area available for electron transfer is increasing and, with progressive nucleation, the number of centres will also increase.

1.10 MULTIPLE ELECTRON TRANSFER

The overall chemical change in many electrode reactions involves the transfer of more than one electron. An extreme example is the complete oxidation of azodyes to carbon dioxide, nitrate and possibly sulfate; such reactions can involve in excess of 100 electrons. It is important to recognize that, in general, electrode reactions occur in single electron steps. Indeed, there is a strong case for writing $n = 1$ in all the equations in Section 1.3.2. Almost always, when the overall chemical change involves more than one electron, a complex mechanism is likely. It is probable that chemical steps either on the surface or in the solution close to the electrode will be separate single $1e^-$ events. This was illustrated in Section 1.7 and will be elaborated further in Chapter 5.

1.11 SUMMARY

The objective of this chapter was to set out the characteristics and physical chemistry of relatively simple electrode reactions and electro-chemical cells. It is only possible to expand the discussion to more complex systems when there is recognition of the complications that can arise and the difficulties and possibilities that each introduces. Hence the preliminary discussion of adsorption, coupled chemical reactions and the deposition of solid phases on electrodes. In later chapters many of these topics will be discussed more thoroughly and illustrated further. It should be emphasized that the same concepts underlie both laboratory studies and electrochemical technology.

Further Reading

1. Southampton Electrochemistry Group, *Instrumental Methods in Electrochemistry*, Ellis Horwood, Chichester, republished 2001.
2. A. J. Bard and L. R. Faulkner, *Electrochemical Methods*, John Wiley & Sons, New York, 2001.
3. K. B. Oldham and J. C. Myland, *Fundamentals of Electrochemical Science*, Academic Press, New York, 1994.
4. *Laboratory Techniques in Electroanalytical Chemistry*, ed. P. T. Kissinger and W. R. Heineman, Marcel Dekker, New York, 1996.
5. Z. Galus, *Fundamentals of Electrochemical Analysis*, Ellis Horwood, Chichester, 1994.
6. J. O'M. Bockris and A. K. N. Reddy, *Modern Electrochemistry, Volume 2A – Fundamentals of Electrodics*, Kluwer Academic, New York, 2001.
7. J. O'M. Bockris, K. N. Reddy and M. Gamboa-Aldeco, *Modern Electrochemistry, Volume 2B – Electrodics in Chemistry, Engineering, Biology and Environmental Science*, Kluwer Academic, New York, 1998.
8. F. C. Walsh, *A First Course in Electrochemical Engineering*, The Electrochemical Consultancy, Romsey, 1993.
9. D. Pletcher and F. C. Walsh, *Industrial Electrochemistry*, Chapman & Hall, London, 1990.

CHAPTER 2

The Two Sides of the Interface

2.1 INTRODUCTION

An essential objective of any course on electrochemistry must be to understand the electron-transfer event at the electrode surface. The electron passes between a delocalized orbital within a metal (the only type of electrode considered in this chapter) and an orbital localized on an ion or molecule in solution. This and the following chapter set the scene for the discussion of electron transfer on a molecular level.

This chapter reviews the structure and chemistry of both the electrode and electrolyte solution. The properties of the reactant and product of electron transfer will be determined by several factors, including solvation, complexation and protonation equilibria and hence these topics will be introduced. In Chapter 8, PEM fuel cells will be discussed and, in such cells, the electrolyte is an ion-conducting polymer; such polymers and membranes fabricated from them are central to several modern electrochemical technologies as well as being useful in the laboratory. Hence, such membranes will also be introduced briefly.

2.2 METALS

2.2.1 Bulk Structure and Properties

A property shared by all metals is a low electrical resistance and this indicates that there are electrons able to move freely through the metal lattice. A good model to understand their high electrical conductivity

A First Course in Electrode Processes, 2nd Edition
By Derek Pletcher
© Derek Pletcher 2009
Published by the Royal Society of Chemistry, www.rsc.org

considers the metal to consist of a highly ordered lattice of the metal ions with the valence electrons delocalized throughout the lattice structure. This model then approximates to a system where a 'sea of electrons' is constrained to exist within a 'box' with the dimensions of the metal phase. In terms of energy levels, the model equates to a large number of closely spaced electronic levels, many filled but also many unfilled at only slightly higher energy. The highest level filled is known as the Fermi level. In fact, the energy levels are so closely spaced that, at ambient temperatures and above, the situation approximates to an infinite number of both filled and empty levels around the Fermi level. Of course, the metal ions in the lattice influence strongly the properties of the electrons within the metal; to form the 'sea of electrons' it is necessary that the atoms have one or more easily ionized electrons and the strength of interaction between ion and electron will determine both the Fermi level and the electrical conductivity of the metal.

Table 2.1 reports the electrical conductivity of some common electrode metals as well as two forms of carbon routinely used as electrodes. It can be seen that, even within the metals, there can be substantial differences in electrical conductivities resulting from difference in binding energies between ions and electrons within the lattice. The reason for preferring copper for the transmission of electricity and for current contacts to cells is clear. While the carbons show 'metal like' behaviour, their electrical conductivities are substantially lower.

Some properties of electrodes can be understood only through knowledge of the bulk structures of metals. The lattices can be envisaged by considering the metal atoms to be hard spheres. Within a single layer, there is only one way to pack spheres of equal diameter. The different lattices result from the packing of layers. Three packing arrangements are common (illustrated in Figure 2.1):

1. In the hexagonal close packed structure, Figure 2.1(a), the basic unit is made up of 17 spheres in a three-tier structure in the form of

Table 2.1 Electrical conductivities, κ, of metals and carbons commonly used as electrode materials.

Electrode Material	$\kappa/S\,cm^{-1}$	Electrode Material	$\kappa/S\,cm^{-1}$
Silver	6.3×10^5	Platinum	1.0×10^5
Copper	6.0×10^5	Lead	4.5×10^4
Aluminium	3.5×10^5	Titanium	1.8×10^4
Zinc	1.8×10^5	Mercury	1.0×10^4
Nickel	1.6×10^5	Graphites	$2-20 \times 10^2$
Iron	1.1×10^5	Vitreous carbons	$1-4 \times 10^2$

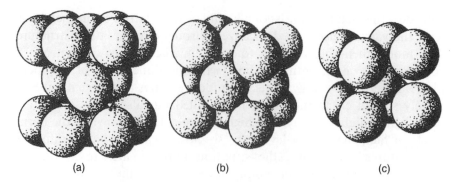

Figure 2.1 Fundamental units for the structure of metals: (*a*) hexagonal closed packed lattice, (*b*) face-centred cubic lattice and (*c*) body-centred cubic lattice.

  a hexagon. The top and bottom layers consist of six spheres surrounding a central one while in the intervening layer, the spheres lie in the cavities between each group of three spheres in the top and bottom layers. Metals with a hexagonal close packed structure include Cd, Co, Hg, Mg and Zn.

2. In the face-centred cubic structure, Figure 2.1(b), the basic unit is a cube with spheres at each corner as well as at the centre of each face. Metals with this structure include Ag, Al, Au, Cu, Ni and Pt.

3. Finally, in the body-centred cubic structure, Figure 2.1(c), the basic unit is again the cube but there are spheres at the corners and at the centre of the cube. Examples of this structure are Cr, Fe, Li and Na.

  In each case the bulk structure is made up of a lattice of these repeated units.

2.2.2 Surface Structure

Throughout this book, the description of simple electron-transfer reactions will assume that the detailed structure of the surface is unimportant. In contrast, any reaction that involves the adsorption of a species from solution and heterogeneous chemical steps on the surface is likely to depend strongly on the surface structure. The separation of metal atoms on the surface determines the possibility of bonding an adsorbate to more than one surface atom, the kinetics of two species on neighbouring atoms reacting together or the probability of forming a bond between two adsorbed species in a concerted reaction.

  In addition to different interatomic arrangements and spacings on a perfect crystal surface, several types of special sites (or defects) influence the surface chemistry. Figure 2.2 defines edge sites, kink sites, edge

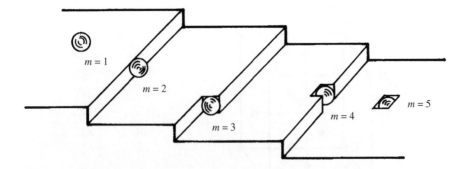

Figure 2.2 Cubic coordination sites in the lattice of a perfect single crystal: (i) surface site, $m=1$; (ii) edge site, $m=2$; (iii) kink site, $m=3$; (iv) edge vacancy, $m=4$; and (v) surface vacancy, $m=5$.

vacancies and surface vacancies where the potential number of neighbouring sites available for multi-site bonding of an adsorbate varies from two to five and such defect sites are often dominant in determining the rate of catalysis and electrocatalysis. Unless great care is taken to prepare a single crystal surface, all electrode surfaces are polycrystalline, consisting of many interlocking crystals. They are not flat on an atomic scale and are likely to have many 'active sites'. These may include the grain boundaries, scratches or other damage during polishing as well as the types of sites shown in Figure 2.2. Another major factor influencing catalytic activity is that many metals will be covered by a surface layer of oxide or other corrosion film unless the preparation of the surface succeeds in removing it and prevents it re-forming. Certainly, it is to be expected that the rate of reactions involving surface chemistry will depend on the history and pretreatment of the electrode surface.

In terms of understanding adsorption and surface chemistry, it can be advantageous to study the electrode reaction on single crystal surfaces. Such surfaces are produced by cutting through the crystal lattice at a defined angle. Then, the surface has a characteristic arrangement of atoms and each atom has a particular number of nearest neighbours with defined positions and known separations. This is illustrated in Figure 2.3 for a metal with a face-centred cubic structure. Figure 2.3 (a) shows another representation of the basic unit where only the positions of the centres of each atom are shown while the table reports low index planes exposed by cutting through a particular set of atoms in the crystal. The various crystal planes are reported using Miller indices (*hkl*). The configuration of surface atoms on three single crystal faces are shown in Figure 2.3 (b) and it is immediately apparent that the local

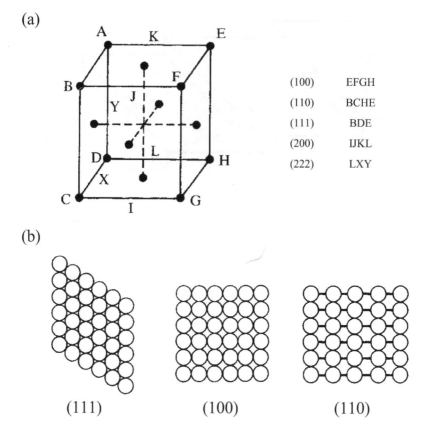

(100)	EFGH
(110)	BCHE
(111)	BDE
(200)	IJKL
(222)	LXY

Figure 2.3 (*a*) Unit cell for a face-centred cubic metal, *e.g.* Pt, and low index planes and (*b*) the arrangement of Pt atoms in three single crystal surface planes.

arrangement of atoms is quite different in each of the three single crystal faces. The number of nearest neighbours and/or the spacing are significantly different.

It is to be expected that each grouping of surface atoms will have a different catalytic activity due to differences in the ability of the surface (a) to adsorb polyatomic molecules, (b) to allow several surface sites to interact with a single adsorbate and (c) to support the efficient reaction of two species adsorbed on neighbouring atoms. These factors, together with the properties of the metal atoms themselves, determine the catalytic activity of single crystal surfaces.

The past 25 years have seen many studies on single crystal electrodes and these certainly confirm that adsorption characteristics and the rate of electrocatalytic reactions are very sensitive to the crystal plane exposed to solution. The literature further shows that the experimental data are very sensitive to contamination and hence the pretreatment of the surface. Other important factors are the extent of adsorption of ions from the electrolyte and surface reconstruction. The latter can result even as a consequence of the experiment being carried out; for example, changes in potential during voltammetry can lead to surface re-construction (hence the response can change substantially and irreversibly during a series of cyclic voltammograms). Certainly, the study of single crystal surfaces requires great care and it is essential to devise an experimental procedure that is reproducible and tested to ensure the correct interpretation of the data. The extent to which electrochemical responses can depend on the crystal plane exposed to solution is, however, illustrated by a typical set of cyclic voltammograms in Figure 2.4. The cyclic voltammograms are for three platinum single crystal electrodes in 0.5 M H_2SO_4; in each case, the surfaces have been flame annealed and cooled in air immediately prior to the experiment. The peaks negative to $+500$ mV *vs* RHE are all due to the adsorption and desorption of hydrogen:

$$H^+ + Pt + e^- \longrightarrow Pt\text{-}H \qquad (2.1)$$

At all surfaces, the responses during the positive and negative going scans are almost mirror images about the zero current axis, indicating that the adsorption and desorption of hydrogen are rapid processes. Also, the charges associated with adsorption/desorption are equal, similar on each of the crystal planes and consistent with the formation and oxidation of a monolayer. In contrast, the shapes of the responses and even the number of peaks are quite different and this reflects differences in the Gibbs free energy of adsorption of hydrogen at specific surface sites (a more positive potential for H adsorption indicates a stronger interaction between a surface site and a hydrogen atom) as well as perhaps differences in the adsorption of bisulfate, HSO_4^-. Notably, the responses are all significantly different from the voltammogram at polycrystalline platinum (Figure 6.5).

This section has looked exclusively at the ways in which the surface properties are related to the metal structure. Of course, the chemistry of the surface is also strongly influenced by the choice of electrolyte medium. Most obviously, the stability of metals to corrosion and the properties of corrosion films on the surface can be very different in different media. For example, some films will support simple electron-transfer reactions while others will totally passivate the surface to further oxidation/reduction. Indeed, the properties of the interface always result from the properties of both the electrode material and the electrolyte as well as the interactions between the two phases. Moreover,

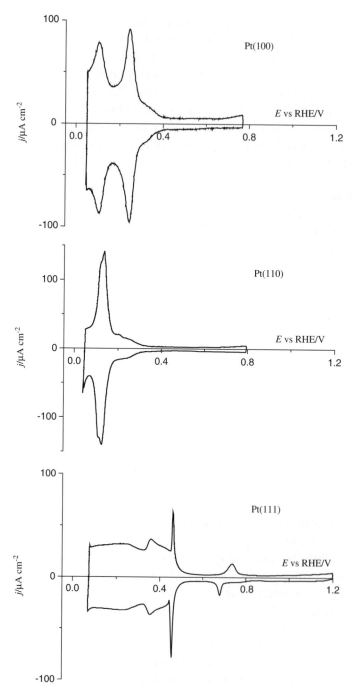

Figure 2.4   Cyclic voltammograms for three single crystal faces of Pt in 0.5 M H_2SO_4. Surfaces were flame annealed and air cooled. Potential scan rate $= 0.1\,V\,s^{-1}$. Data taken from J. Clavilier in *Interfacial Electrochemistry – Theory, Experiment and Applications*, ed. A. Wieckowski, Marcel Dekker, New York, 1999. The response for polycrystalline Pt is shown in Figure 6.5.

the properties are a strong function of potential and also vary with parameters such as temperature.

2.3   ELECTROLYTE SOLUTION

The solutions for most electrochemical experiments contain a solvent, a high concentration of an electrolyte and the electroactive species.

2.3.1   Solvent

A wide range of solvents can be used for electrochemical experiments in the laboratory and also find some application in technology. The main properties required are:

- to be liquid at the temperature of the experiment;
- the ability to dissolve salts to give solutions that have ionic conductivity;
- stability to oxidation/reduction at the electrode, at least in the potential range of interest for the experiment/application;
- stability to the acid/base commonly formed at the anode/cathode;
- non-interference with the intended chemistry at the electrode, *e.g.* by reaction with intermediates;
- to provide a facile and clean counter electrode reaction – an advantage of water as a solvent is that H_2 and O_2 evolution are such reactions.

The ability of a solvent to dissolve electrolytes to give conducting solutions can be assessed using any one of several thermodynamic solvation parameters or by a consideration of the donor/acceptor properties of the solvent. Dielectric constants are, however, readily available and are a useful guide; a dielectric constant of < 10 indicates a poor solvent for electrochemistry while a value of > 30 can be taken as an indication that high quality experiments should be possible. The solubility of simple salts is another useful guide. Table 2.2 gives these properties for some typical solvents. With organic solvents, particularly those with a low dielectric constant, tetraalkylammonium salts with a large organic cation are often the preferred electrolytes.

Water is, however, the most common solvent for electrochemistry and it is important to recognize that it is a solvent with unique properties favourable to electrochemistry. Firstly, unlike most liquids, which may

Table 2.2   Some properties of solvents routinely used in electrochemistry.

Solvent	Liquid range/K	Dielectric constant	Solubility/g kg⁻¹ NaBr	NaCl
Water	273–373	80	940	360
Acetic acid	290–391	6	11	1.4
Methanol	179–338	33	174	14
Acetonitrile	28–355	38	0.4	0.003
Dimethylformamide	212–426	37	103	0.4
Dimethyl sulfoxide	291–462	47	57	4.7
Propylene carbonate	224–515	69	8	Insoluble
Tetrahydrofuran	208–340	7	Insoluble	Insoluble
Methylene dichloride	176–313	9	Insoluble	Insoluble

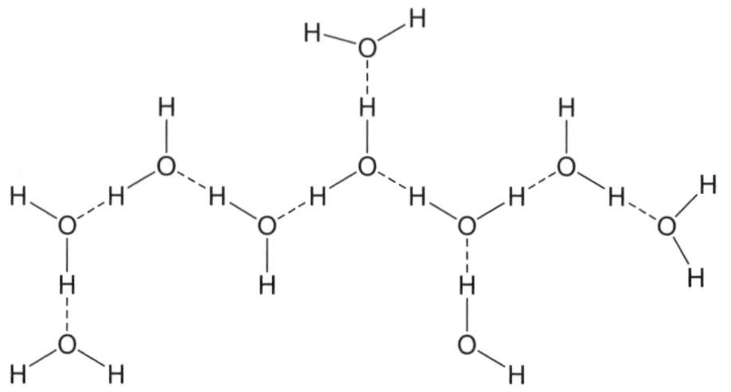

Figure 2.5   Structure of a water oligomer in liquid water.

be pictured as a collection of randomly orientated molecules with weak intermolecular interactions, water is a solvent with considerable structure resulting from the ability of each water molecule to form hydrogen bonds with its neighbours. In consequence, liquid water has an extensive structure with many oligomers and polymers (Figure 2.5); this is, however, not a rigid structure because the hydrogen bonds are continuously breaking and new ones forming. Secondly, water is a small molecule with a substantial dipole moment. Hence, it is readily polarized and can interact electrostatically with charged species. In other words, it solvates ions readily and this leads to the facile dissolution of salts to form conducting solutions. Cations solvate more strongly than anions; the Gibbs free energies of hydration of cations such as H^+, Li^+, and Na^+ are very large and, typically, such ions have an inner hydration shell containing six water molecules, with the oxygen atoms of the water donating electrons to the cation, and an outer hydration shell consisting of more loosely bound water molecules. In general, the water dipole can interact less strongly with anions but the molecular forces are often supplemented by hydrogen bonding, *e.g.* to oxygen atoms in an anion. Finally, water self-ionizes to a small extent; even in neutral solutions there is a low concentration of H^+ and OH^- ions ($\sim 10^{-7}$ M). But, more importantly, water can act as both a proton donor and a proton acceptor and, in general, acid/base equilibria in aqueous solution are facile and rapid. It is the combination of these properties that makes water an excellent solvent for acids, bases and salts, leading to the formation of ions and to highly conducting solutions.

In most situations, it is also essential that the electroactive species is soluble to the required extent. In Chapter 1, it was emphasized that the maximum rate of an electrode reaction is determined by a mass transport limitation, and in the mass transfer regime in typical electrolytic cells the limiting current density for a 1 M solution of reactant is <250 mA cm^{-2}. Hence, applications that require a significant current density are dependent on the reactant being highly soluble or on the use of a more sophisticated approach (*e.g.* an indirect electrolysis, two-phase electrolysis, a three-dimensional electrode). As a general rule, solutes are most soluble in solvents of similar polarity, with large enhancements possible if Lewis acid–Lewis base interactions are possible. Organic molecules tend to be most soluble in organic solvents, where it is difficult to find salts that are soluble and ionize to give conducting solutions. Hence, it is often necessary to compromise by using water/organic solvent mixtures or a dipolar aprotic solvent. The dipolar aprotic solvents have an organic backbone with polar substituents and dissolve a wide range of ionic and neutral species. Even then, to obtain a high concentration of an inert ion in solution, it is often necessary to select a salt with a large organic cation, with tetrabutylammonium tetrafluoroborate being a common choice.

2.3.2 Electrolyte

The electrolyte has several roles in practical electrochemistry. Most importantly, it reduces the resistance of the solution between electrodes. It also (a) avoids migration having a significant role in the mass transport of the reactant and/or the product of electron transfer, (b) provides the ions to form a simple double layer, (c) can simplify the discussion of non-ideal solutions and (d) acts as a pH buffer, particularly within the reaction layer at the electrode surface. These roles are discussed in more detail in Chapter 6.

To perform these roles, the electrolyte must ionize to form cations and anions in the solution. Hence, strong electrolytes, *i.e.* electrolytes that ionize fully, are preferred. Weak electrolytes do not ionize fully; an example is acetic acid:

$$CH_3COOH \rightleftharpoons CH_3COO^- + H^+ \tag{2.2}$$

and the degree of ionization is discussed in terms of a dissociation constant:

$$K_{dis} = \frac{c_{H^+} c_{CH_3COO^-}}{c_{CH_3COOH}} \tag{2.3}$$

Much of our knowledge of electrolytes in solution comes from studies of electrolytic conductivity, usually calculated from a measurement of the resistance of the solution between two electrodes using the expression:

$$\kappa = \frac{S}{AR} \tag{2.4}$$

where κ is the electrolytic conductivity, S the separation of the electrodes, A the electrode areas and R the resistance of the solution. The resistance should be measured with a high frequency AC instrument to avoid artefacts associated with the electrodes.

Table 2.3 reports some typical electrolytic conductivities for aqueous solutions. It can be seen that the values are substantially lower than those reported for the metals in Table 2.1 and the electrolytic conductivities decrease in the order: acids > bases > neutral strong electrolytes > weak electrolytes.

The high values for acids and bases result from a special mechanism (the Grotthuss mechanism) for the transport of H^+ and OH^- ions in aqueous solutions. This involves protonation and deprotonation at the opposite ends of the water oligomers with simultaneous rearrangement

Table 2.3 Ionic conductivities, $\kappa/S\,cm^{-1}$, as a function of concentration for some acids, bases and salts at 298 K. Data taken from *Handbook of Electrolyte Solutions*, ed. V.M.M. Lobo, Elsevier, Amsterdam, 1989.

	Concentration/M			
	10^{-3}	10^{-2}	0.1	1.0
H_2SO_4	7×10^{-4}	5.40×10^{-3}	0.056	0.63
HCl	4.2×10^{-4}	4.10×10^{-3}	0.039	0.33
NaCl	1.2×10^{-4}	1.18×10^{-3}	0.011	0.09
Na_2SO_4	1.3×10^{-4}	2.02×10^{-3}	0.016	0.06
KCl	1.4×10^{-4}	1.41×10^{-3}	0.013	0.11
$MgCl_2$	1.2×10^{-4}	1.15×10^{-3}	0.012	0.12
$CuCl_2$	2.4×10^{-4}	2.20×10^{-3}	0.018	0.10
NaOH	2.4×10^{-4}	2.49×10^{-3}	0.022	0.19
KOH	2.7×10^{-4}	2.65×10^{-3}	0.025	0.21

of oxygen to hydrogen bonds within the oligomer, leading to net movement of H^+ or OH^- ions from one end of the oligomer to the other. The decrease in electrolytic conductivity for the same concentration of chloride salt along the series $KCl > NaCl > LiCl$ reflects the differences in the sizes of the hydrated cations; although the ionic radius in the gaseous state of Li^+ is smaller than Na^+, Li^+ attracts more waters of hydration than Na^+ and K^+ and the larger hydrated cation moves more slowly through the solution. With strong electrolytes, the electrolytic conductivity is proportional to the electrolyte concentration, at least up to highly concentrated solutions. The mobility of ions is largely determined by the size of the ion and the viscosity of the medium and at very high concentrations the viscosity increases markedly. The electrolytic conductivities of non-aqueous solutions are invariably substantially below those for aqueous solutions.

Transport numbers, t_+ and t_-, define the fraction of charge carried through the solution by the cation and anion, respectively. For a solution of a single electrolyte, $t_+ + t_- = 1$, and usually in neutral aqueous solutions $t_- > t_+$ because the anion is smaller than the cation in solution and therefore moves more rapidly through the solution.

For many situations in chemistry, including concentrated solutions of electrolytes, it is advisable to consider deviations from ideal solution behaviour. Such deviations are correctly handled by using 'activities' rather than 'concentrations'. For practical purposes, the activity can be thought of as the 'effective concentration'. In other words, it is a concentration that takes into account the local circumstances of the species and hence reflects more closely the chemistry, both the thermodynamics and kinetics, of the species. In the case of ions in concentrated electrolyte solutions it is the interaction with ions of opposite charge that is the major factor modifying the chemistry. For each ion in solution, the activity, a, is defined by:

$$a_+ = \gamma_+ m_+ \quad \text{and} \quad a_- = \gamma_- m_- \tag{2.5}$$

where γ is here the activity coefficient and m the molality of the ion (subscripted to indicate cation and anion). For ideal solutions the activity coefficients are 1 and the activities and the concentrations are equal. Since the main deviations in concentrated electrolyte solutions arise from mutual interactions of the ions of opposite charge, it is not possible to separate the effects due to cations or anions alone. Therefore, the system has to be discussed in terms of mean ionic activity coefficients and mean ionic activities, defined for a 1 : 1 electrolyte by:

$$\gamma_\pm^2 = \gamma_+ \gamma_- \quad \text{and} \quad a_\pm^2 = a_+ a_- \tag{2.6}$$

The mean ionic activity coefficient, γ_+, describes the deviation of electrolyte in solution from its standard state, *i.e.* the hypothetical situation where the electrolyte exists at unit molality but has the environment of an infinitely dilute solution – there are no ionic interactions. Debye and Hückel were the first to develop an accepted description of relatively dilute electrolyte solutions that took into account the electrostatic interactions and repulsions between the ions in solution. The main result is an equation that allows an estimation of the mean ionic activity coefficient:

$$\log \gamma_{\pm} = -0.509|z_+z_-|I^{1/2} \tag{2.7}$$

where z_+ and z_- are the charges on the ions and I is the ionic strength, defined by:

$$I = \frac{1}{2}\sum_i m_i z_i^2 \tag{2.8}$$

Equation (2.7) treats all ions of like charge as equivalent and hence does not take their chemistry into account. However, it does recognize that electrolytes with multivalent ions will deviate more strongly than 1:1 electrolytes (I depends on the square of the charge) because of the stronger effect of multi-charged ions on the electrostatics and their greater tendency to form ion pairs. Moreover, for solutions containing more than one electrolyte, the ionic strength contains a term for each ion formed in solution and concentrated electrolytes will have a strong effect on the mean ionic activity coefficient for electrolytes present in low concentrations (including electroactive species and products). This causes the activity coefficient to be similar for all ions present in solutions containing a high concentration of another electrolyte. Note that this simplifies discussion of couples such as Fe^{3+}/Fe^{2+} because the activity coefficients for Fe^{3+} and Fe^{2+} in the Nernst equation will then cancel and it is safe to use concentrations rather than activities. This simplification will not, however, apply to couples such as Cu^{2+}/Cu. Since the early work of Debye and Hückel, several equations more complex than (2.8) have been proposed to extend the concepts to more concentrated electrolyte solutions. Here, it is more important to recognize the magnitude of errors possible by ignoring non-ideality and this is best achieved by looking at experimental values of activity coefficients (Table 2.4). Clearly, the activity coefficients fall substantially below 1 as the concentration of electrolyte increases – ionic interactions are stabilizing the ions, causing marked deviations from ideal behaviour. As expected, the deviations become greater when multivalent ions are involved, but for 1 mol kg^{-1} solutions the error caused by using

Table 2.4 Mean ionic activity coefficients as a function of concentration for some electrolytes in aqueous solution at 298 K. Data taken from *Handbook of Electrolyte Solutions*, ed. V.M.M. Lobo, Elsevier, Amsterdam, 1989.

	Electrolyte			
$m/mol\,kg^{-1}$	*NaCl*	*KCl*	*NaOH*	H_2SO_4
10^{-3}	0.966	0.965	–	0.830
10^{-2}	0.904	0.901	0.861	0.544
0.1	0.780	0.769	0.780	0.265
1.0	0.660	0.606	0.680	0.130
2.0	0.670	0.576	0.700	0124
4.0	0.780	0.579	0.890	0.171

concentrations rather than activities is always $> 30\%$. It can also be seen that the experimental activity coefficients pass through a minimum at high concentrations; this unexpected result is thought to arise because of a change in the structure of water due to the fraction of the solvent tied up by the ions.

2.3.3 Reactant, Intermediates and Product

In the electrode reaction:

$$O_{soln} + ne^- \rightleftharpoons R_{soln} \qquad (2.9)$$

both the thermodynamics and kinetics of the overall reaction are determined by the chemistry of the oxidized and reduced species in solution. It is immediately clear that:

1. Any change to the solution composition that leads to thermodynamic stabilization of the species O will make the reduction more difficult (as well as the reverse reaction easier) and lead to a negative shift in the formal potential for the couple. Conversely, any change to the solution composition that leads to thermodynamic stabilization of the species R will make reduction easier (and oxidation more difficult), hence causing a positive shift in the formal potential.
2. During the electrode reaction the electroactive species must change structure to that of the product. One of the key factors determining the kinetics of the process will be the similarity in the structure of O and R. For example, the need for bond breakage, changes to the number of ligands or major changes to the geometry of the species

are likely to require a large activation energy and hence lead to a slow reaction. In contrast, when O and R have similar structures the electron-transfer reaction is likely to be fast.

Hence, one purpose of this section is to review the chemistry of species in solution and consider the way in which it influences electrode reactions. In any electrode reaction, it is inevitable that either the reactant or a product is an ion; it is possible that (a) O and R are both cations, (b) O is a cation and R a neutral molecule, (c) O is a neutral molecule and R an anion, (d) both O and R are anions and (e) O and R are both neutral but other ions, usually H^+ or OH^-, are formed. Hence, the discussion must cover both ions and neutral molecules. In many systems, both O and R must be soluble in the electrolyte solution and this implies that there are interactions between solutes and solvents. As a generalization, species are solvated most strongly when the solute and the solvent have similar polarities. Thus, salts are most soluble in solvents whose molecules have strong dipoles; water is the best example. Conversely, non-polar, neutral molecules (*i.e.* most organic compounds) are poorly soluble in water but dissolve readily in non-polar organic solvents (*e.g.* diethyl ether, toluene, hexane). As a result, experiments or applications requiring a concentrated solution of an organic compound require a compromise, a water/organic solvent mixture or an aprotic solvent (*e.g.* acetonitrile, dimethylformamide) to ensure that both ions and neutral molecules have sufficient solubility.

Diffusion is an important mode of mass transport for the reactant and product of the electrode reaction. It should be recognized that species are always moving around in solution. In bulk solution, the movement is random in nature and, in the absence of a concentration gradient, there will be no net change in the distribution of the species in solution. Net transport arises where there is a concentration gradient, see Chapter 1, and such gradients always occur close to a surface where chemical change is occurring. Whether considering systems that involve diffusion only or both diffusion and convection, it can be seen from the equations in Chapter 1 that the rate of mass transport depends on the diffusion coefficient, D, for the transported species. The diffusion coefficient is given by:

$$D = \frac{kT}{6\pi\eta r_{\text{eff}}} \tag{2.10}$$

where here η is the viscosity of the medium and r_{eff} is the radius of a sphere equivalent to the dimensions of the diffusing species in solution. The latter is not a well-defined quantity but, clearly, the value of the diffusion coefficient is inversely dependent on the size of the diffusing species in the electrolyte medium and the viscosity of the medium. The size of the diffusing species will depend on the extent of solvation and interaction with other species in the medium; even the relative size of species in the electrolyte medium can be very different from their nominal sizes in the gas phase. Hence, it is often difficult to predict the relative values of diffusion coefficients. In contrast, the dependence of the diffusion coefficient on viscosity is straightforward and generally follows Equation (2.10). In aqueous solutions, diffusion coefficients normally fall in the range 0.5–$1.0 \times 10^{-5}\,\mathrm{cm^2\,s^{-1}}$, with only species such as the proton (because of the Grotthuss mechanism for transport through solution) and oxygen (a very small species) giving higher values and very large molecules or polymers giving smaller values.

2.3.3.1 Transition and Heavy Metal Species. In many experiments/applications, the electroactive species is a transition metal entity, and transition metals exist in solution as several distinct types of species.

The transition metal may be present as a solvated cation. In aqueous solution, it is normal for the cation to be surrounded by a defined number, commonly six, of tightly bound water molecules in an inner solvation shell with further more loosely bound water molecules forming outer solvation shells. Hydrated cations can be much stronger acids than water itself and equilibria of the type:

$$M(H_2O)_6^{n+} \rightleftharpoons MOH(H_2O)_5^{(n-1)+} + H^+ \qquad (2.11)$$

are common. Only in acid solution are the simple hydrated cations stable; as the pH is raised, hydroxyl complexes form stepwise and eventually a species of the type $M(OH)_n$ will precipitate. The pH where the hydroxyl species form depends on the transition metal and its oxidation state – higher oxidation states are more prone to hydrolysis. Because of this difference, only in acid solution do the oxidized and reduced species have very similar structures and then simple, rapid electron transfer is possible. Less is known about the solvation of transition metal ions in other solvents but it is probable that inner and outer solvation shells are again formed. The interaction between the ions and organic solvent molecules is, however, likely to be much weaker than the ion–water bond. Indeed, because of the strong affinity between water and transition metal cations, experiments in non-aqueous solvents are often particularly affected by trace contamination by water.

The transition metal cation may form ion pairs, $M(H_2O)_6{}^{n+}X^-$, with anions in the solution. In an ion pair, the cation and anion retain their full hydration shells and the interaction is electrostatic. Such interactions are common in the concentrated electrolyte solutions used in many electrochemical systems. As noted above, one approach to discussing ion pairing is to consider it as a deviation from ideality leading to $\gamma_+ < 1$.

Many ligands form labile complexes with transition metal ions, *i.e.* there are a series of equilibria (written for an uncharged ligand) for each oxidation state:

$$M(H_2O)_6{}^{n+} + L \rightleftharpoons M(H_2O)_5L^{n+} + H_2O \tag{2.12}$$

$$M(H_2O)_5L^{n+} + L \rightleftharpoons M(H_2O)_4L_2{}^{n+} + H_2O \tag{2.13}$$

substitution continuing to

$$M(H_2O)L_5^{n+} + L \rightleftharpoons ML_6{}^{n+} + H_2O \tag{2.14}$$

In general, L is a ligand that could be neutral or an anion. The extent of complexation and the ratio of the species in solution depend strongly on the concentration of the ligand and may be calculated if the stability constants for the metal/ligand combination are known. It is, however, important to recognize that the speciation results from a competition between the ligand and water (and, if present, other species in the medium) for bonding sites on the transition metal ion. In addition, the solution pH will influence the chemistry of both the cation and the ligand (through protonation and deprotonation equilibria). Although there are exceptions [*e.g.* Cu(I) interacts more strongly than Cu(II) with halide ion and amines], it is usually found that ligands interact more strongly with higher oxidation states. Hence, the addition of ligands to a solution usually makes the reduction of O to R more difficult so that the formal potential shifts negative and moves more negative as the concentration of ligand is increased. The shift in the formal potential with ligand concentration provides a method to study complex formation. For the electrode reaction:

$$ML_p^{(m-pb)+} + ne^- \rightleftharpoons ML_q^{(m-n-qb)+} + (p-q)L^{b-} \tag{2.15}$$

where for simplicity the water ligands have been omitted, the shift in the formal potential on addition of ligand is given by:

$$(E_e^\circ)_L - (E_e^\circ)_{H_2O} = \frac{2.3RT}{nF}\left(\log\frac{K_O}{K_R} - (p-q)\log c_L\right) \tag{2.16}$$

where $(E_e^{\circ})_L$ and $(E_e^{\circ})_{H_2O}$ are the formal potentials in the absence and presence of ligand, respectively, with a concentration, c_L, and K_O and K_R are the overall stability constants for the reactions:

$$M^{m+} + pL^{b-} \rightleftharpoons ML^{(m-pb)+} \tag{2.17}$$

$$M^{(m-n)+} + qL^b \rightleftharpoons ML^{(m-n-qb)+} \tag{2.18}$$

defined by:

$$K_O = \frac{[ML_p^{(m-pb)+}]}{[M^{m+}][L^{b-}]^p} \tag{2.19}$$

$$K_R = \frac{[ML_q^{(m-n-qb)+}]}{[M^{(m-n)+}][L^{b-}]^q} \tag{2.20}$$

It can be seen that a plot of the formal potential *versus* log c_L should be linear (for reversible couples, the formal potential can be read directly from the experimental responses). The slope allows the determination of $(p - q)$ and the intercept an estimation of the ratio of the stability constants K_O/K_R.

The equations are much simpler for electrode reactions such as:

$$Cd(NH_3)_p^{2+} + Hg + 2e^- \rightleftharpoons CdHg + pNH_3 \tag{2.21}$$

where complex formation is a consideration for only one oxidation state since the reduced species is cadmium amalgam where the Cd metal does not interact with ammonia in the electrolyte. Figure 2.6 reports a plot of the formal potential for the Cd(II)/CdHg couple (in fact, the half-wave potential from the *I vs E* curve for the reduction of Cd(II) at a dropping mercury electrode) *versus* the concentration of ammonia. The slope of the plot is $1/(120\,mV)$ and, since $n=2$, $p=4$, cadmium(II) exists in solution as the species $Cd(NH_3)_4^{2+}$. Moreover, since the formal potential in the absence of ammonia is $-580\,mV$ *vs* SCE, the stability constant for the complex is $\sim 10^7$.

The transition metal may also be part of a non-labile complex. For example, in ferricyanide and ferrocyanide, the Fe(III) and Fe(II) are surrounded octahedrally by six tightly bound cyanide ions. Under a wide range of conditions, even in the absence of free cyanide, there is no tendency to dissociate cyanide ligands. Similarly, in ferrocene and the ferrocenium ion, the iron centres are bound strongly to two cyclopentadienyl ligands in a very stable structure. In such non-labile complexes the solvent is present only as an outer solvation shell. The kinetics

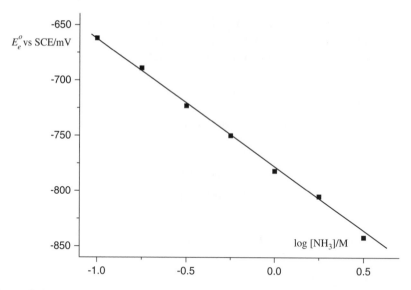

Figure 2.6 Formal potential for the Cd(ıı)/CdHg couple as a function of the concentration of ammonia in a 0.1 M KNO_3 solution.

of reactions involving non-labile complexes are often rapid because there is no major change in structure during the redox process. In addition, ligands forming non-labile complexes are often able to stabilize unusual oxidation states; for example, the electrochemistry of Cp_2Co shows both oxidation and reduction processes and both Co(ııı) and Co(ı) are stable oxidation states. Many macrocyclic ligands form non-labile complexes and also stabilize unusual oxidation states.

2.3.3.2 Organic Molecules.   Most commonly, organic compounds are neutral molecules with limited solubility in water but are highly soluble in organic solvents. Their susceptibility to oxidation and reduction is determined by the energies of the highest filled and lowest unfilled molecular orbital, respectively. The electrochemistry of organic molecules is, however, also strongly influenced by protonation equilibria. This is illustrated in Figure 2.7, which shows a 'scheme of squares' for the $2e^-$ reduction of an organic molecule, A. Here, A could, for example, be an aromatic ketone and the 'scheme of squares' emphasizes the number of potential pathways involving the addition of two electrons and two protons to form the final product (an alcohol if the reactant is a ketone). The pathway will depend on the acidity of the medium and the basicity of both the electroactive species and the intermediates in the reaction. In general, the more reduced species is the

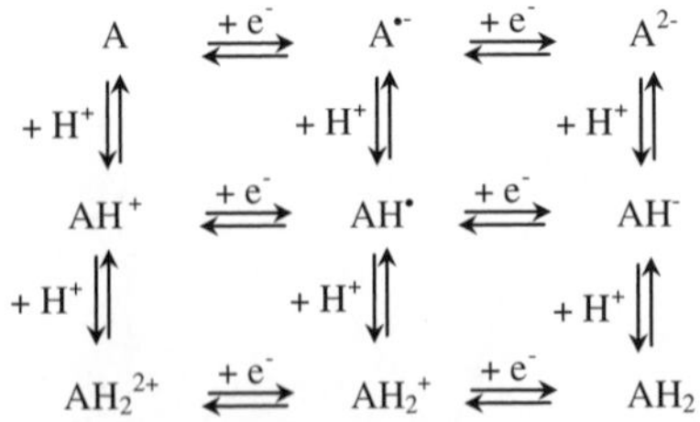

$$A \xrightleftharpoons{+e^-} A^{\bullet-} \xrightleftharpoons{+e^-} A^{2-}$$

$$+H^+ \updownarrow \qquad +H^+ \updownarrow \qquad +H^+ \updownarrow$$

$$AH^+ \xrightleftharpoons{+e^-} AH^{\bullet} \xrightleftharpoons{+e^-} AH^-$$

$$+H^+ \updownarrow \qquad +H^+ \updownarrow \qquad +H^+ \updownarrow$$

$$AH_2^{2+} \xrightleftharpoons{+e^-} AH_2^+ \xrightleftharpoons{+e^-} AH_2$$

Figure 2.7 Scheme of squares; the $2e^-$ reduction of the organic molecule occurs by a sequence of electron transfers and protonations, where six pathways are possible. The order depends on the acidity of the medium and the basicity of A and the reaction intermediates.

stronger base and the protonated species reduce more readily than the unprotonated form. Thus in an aprotic solvent or strongly basic aqueous medium, the first step will be electron transfer to give an anion radical and the anion radical may be a stable species in the medium. If so, the next step may well be the addition of a second electron (perhaps at a more negative potential) and the final product is formed by an eecc mechanism.

In contrast, in strong aqueous acid, the starting material may well protonate or, at least, be partially protonated and, since it is likely that the protonated species reduces more readily, the reduction will occur by a mechanism initiated by a ce sequence. In media of intermediate acidity, the anion radical but not the neutral starting material may protonate and the reduction will be an ece process.

If we consider the simplified scheme:

$$A + H^+ \rightleftharpoons AH^+ \tag{2.22}$$

$$AH^+ + n_1 e^- \longrightarrow P_1 (\text{at } E_1) \tag{2.23}$$

$$A + n_2 e^- \longrightarrow P_2 (\text{at } E_2, \text{ more negative than } E_1) \tag{1.24}$$

it is possible to define three pH regions (dependent on the pK_B of the compound A), see Figure 2.8, where the variation of the reduction potential with pH is different:

1. In very acid solution, AH^+ will be the dominant species throughout the solution and the formal potential will then be E_1, independent of pH, and the product will be P_1.
2. As the pH is increased and approaches the pK_B for A, the bulk solution will contain both AH^+ and A but only AH^+ will reduce at a potential close to E_1. As AH^+ is depleted at the electrode surface,

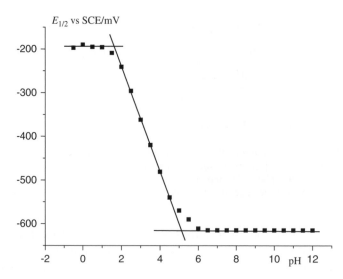

Figure 2.8 Variation of the half-wave potential (formal potential) for the reduction of an organic molecule that is a weak base.

the equilibrium concentration of AH^+ within the reaction layer will be re-established by further protonation of A. Effectively, A will be reduced by a ce mechanism consisting of protonation followed by electron transfer. In this situation, the formal potential will shift negative with increasing pH, in the general case by $2.3\,mRT/F$ mV per pH unit where m protons are involved prior to the electron transfer. The product remains P_1.

3. On further increasing the pH, one reaches the stage where E_1 is no longer positive to E_2 and then unprotonated A will reduce directly. The formal potential will again be independent of pH at E_2 and the product will change to P_2.

   If the protonation step is rapid (usually the case in aqueous solution), a single reduction step will always be observed and the limiting current is a function of pH only if $n_1 \neq n_2$. The limiting current will be mass transfer controlled with respect to all the compound A in solution. In contrast, if the protonation step is slow, two waves can be seen at E_1 and E_2 and the ratio in limiting currents will depend on the kinetics of protonation as well as n_1/n_2. Figure 2.9 shows a specific example of a change in product with pH; pinacol is formed in acid solution and the alcohol in strongly alkaline media.

   In all media, control of pH is important in determining the mechanism and products. Almost all reductions of neutral compounds lead to the formation of base and oxidations to the formation of acid.

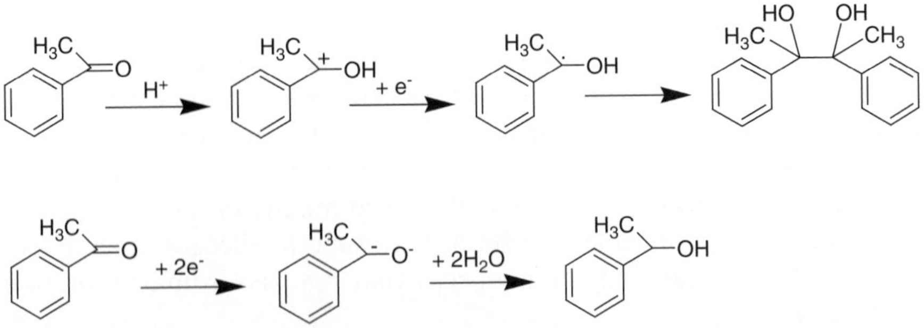

Figure 2.9 Reduction of acetophenone in acidic and alkaline aqueous solution.

Hence, synthetic reactions demand careful choice and control of the pH. For example, in the commercial production of adiponitrile, the electrolyte must be neutral and remain neutral. The desired reaction is:

$$2CH_2=CHCN + 2H_2O + 2e^- \longrightarrow (CH_2CH_2CN)_2 + 2OH^- \qquad (2.25)$$

If the OH^- is not rapidly dispersed from the reaction layer at the electrode surface, base-catalysed addition of water to the acrylonitrile decreases the selectivity of the reaction:

$$CH_2=CHCN + H_2O \xrightarrow{OH^-} HOCH_2CH_2CH \xrightarrow{OH^-} O(CH_2CH_2CN)_2 \qquad (2.26)$$

However, it is not possible to make the catholyte acidic because then the electrode reaction changes to:

$$CH_2=CHCN + 2H^+ + 2e^- \longrightarrow CH_3CH_2CN \qquad (2.27)$$

2.3.3.3 *Other Electroactive Species*. Many other types of inorganic species, including H_2O, O_2, CO_2, SO_2, H^+, Cl^-, Br^- and NO_3^-, are electroactive under appropriate conditions but it is beyond the scope of this book to discuss all their solution chemistries. Notably, pH is again the major experimental parameter influencing their electrochemistry. The electrolyte pH determines whether the protonated reactant is the major species in bulk solution or whether protonation occurs within the reaction layer at the surface prior to electron transfer. In addition, for species such as Cl^- and Br^-, there is a wide range of coupled chemistry that varies with pH and the pH determines the product.

2.4 ION-PERMEABLE MEMBRANES

A separator is often essential to the success of an experiment/application involving an electrolysis cell. It may be required to separate the reactants at the two electrodes (*e.g.* for safety reasons as in an oxygen/hydrogen fuel cell), to separate the products (likely to react chemically in solution if allowed to mix) or to prevent the back reaction at the counter electrode. Commonly, in the laboratory the separator may be a glass frit or microporous polymer sheet but such structures will introduce no selectivity in the transport and merely retard mixing of the solutions on either side. Ion-permeable membranes are superior because they introduce substantial selectivity in the species that can pass from one solution to the other.

An ion-permeable membrane (often called an ion-exchange membrane) is a thin sheet of a polymer designed to allow the transport either of cations or anions but not both. Transport of ions should occur only by migration; ideally, the transport number for the ion of interest should be one and the membrane polymer should resist the transport of solvent and other neutral molecules. The polymers will have fixed ionic groups (usually sulfonate or carboxylate for cation-permeable membranes or tetraalkylammonium groups for anion-permeable membranes) that are designed to prevent the passage of ions of the same charge and allow the transport of ions of opposite charge. The fixed ionic groups are bound to a polymer background designed to allow the appropriate level of hydration. Hydration is essential to obtain an acceptable conductivity but too much water within the structure will diminish the selectivity of ion transport and even cause the polymer to dissolve. The most successful cation-permeable membranes are based on a perfluorohydrocarbon backbone with sulfonate or carboxylate fixed ionic groups. Figure 2.10 shows a typical chemical structure. The perfluorohydrocarbon backbone provides rigidity and resistance to chemical attack by acid, alkali and redox reagents while the sulfonate groups introduce the selectivity in ion migration. The balance between the hydrophobic backbone and the hydrophilic fixed ionic groups provides the mechanism to control the

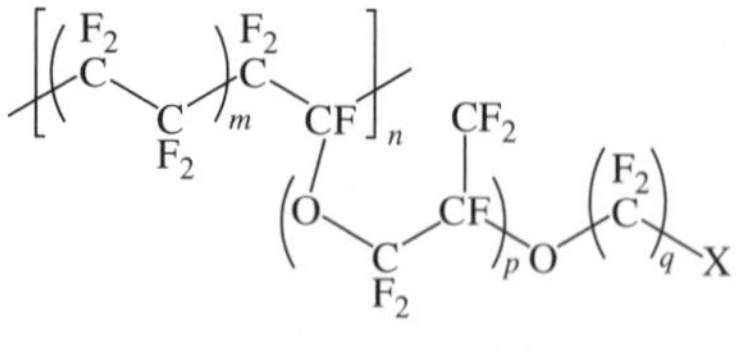

$$X = -SO_3^- \text{ or } -COO^-$$
$$m = 5 - 15, n = 600 - 1500,$$
$$p = 1 - 3, q = 1 - 5$$

Figure 2.10   Chemical structure of perfluorinated polymers (*e.g.* Nafion™) used in the fabrication of cation-permeable membranes.

water level. A frequently used model for these membranes pictures the polymer as having large areas of the hydrophobic, perfluorinated hydrocarbon chains with the hydrophilic sulfonate groups clustered together. When contacted by aqueous solutions, the fixed ionic groups and the mobile counter cations hydrate and attract further water, leading to the formation of many, contorted molecular scale channels flooded with water through which the cations migrate.

During electrolysis, a potential gradient is created between the electrodes and, with a cation-permeable membrane, cations will migrate from the anolyte to the catholyte. This gradient also provides a driving force for anions to migrate through the membrane from catholyte to anolyte. Hence, the structure of the cation-permeable membrane has to prevent this unwanted process. Modern membranes give good selectivity with $t_+ = 0.95-0.99$ but are never totally selective for cations *vs* anions. To introduce selectivity between cations is much more difficult and the electrolytic cell is usually operated under conditions where there is only one cation in the anolyte so that selectivity between cations is not an issue. In addition, membranes allow the diffusion of neutral molecules at a finite rate in both directions. Water is a particular problem; it moves through the membrane with the cations (as water of hydration – typically >6 moles of H_2O will be transported with each mole of cation), due to pressure differences across the membrane and also because of differences in activity if there are different solutions on the two sides of the membrane.

Since the transport of ions through the membrane is by migration, a voltage drop across the membrane is a physical necessity. In practice, the voltage drop needs to be minimized since it contributes to the energy consumption of the cell. Hence the electrical resistance of the membrane should be minimized; this is achieved by employing a polymer with a high ionic conductivity and fabricating the membrane as thin as possible (compatible with mechanical stability). Modern membranes are typically 30–100 μm thick and have an area resistance in the range 0.2–1.0 ohm cm^2. Therefore, with a current density in the range 0.1–1.0 A cm^{-2}, the voltage drop is 0.02–1.0 V, which is comparable to the voltage drop across typical electrolyte compartments. It should also be recognized that the flux of ions through the solution to the membrane surface must be higher than the flux of ions through the membrane (otherwise the membrane will come under mass transport control and the voltage drop will rise steeply). Hence for high current densities, the mass transport to the membrane must be efficient. The situation at the membrane surface can be discussed in terms of a boundary layer in the same way as mass transport to an electrode (Chapter 1).

Table 2.5 Properties important in the application of ion-permeable membranes in electrochemical technology.

- High ionic conductivity (to give a low *IR* drop when current is passed) but negligible electronic conductivity at the cell operating temperature
- Selective transport of one ion, $t_+ = 1$
- Low transport of solvent and other neutral molecules
- Chemical stability to reactants, products and electrolyte at the cell operating temperature
- Mechanical stability and compatibility with gaskets
- Maintains performance at high current density

Table 2.5 lists the desirable properties for the membrane. Membranes never meet the requirements entirely and hence it is important to operate them under conditions where their shortcomings are not critical to the application. Membranes will be discussed further in Chapter 8 on PME fuel cells.

Further Reading

1. J. O'M. Bockris and A. K. N. Reddy, *Modern Electrochemistry 1, Ionics*, Kluwer Academic, New York, 1998.
2. J. Koryta, J. Dvorak and L. Kavan, *Principles of Electrochemistry*, John Wiley & Sons, New York, 1993.
3. L. Meites, *Polarographic Techniques*, Interscience, New York, 1965.
4. T. A. Davis, J. D. Genders and D. Pletcher, *A First Course in Ion Permeable Membranes*, The Electrochemical Consultancy, Romsey, 1997.

CHAPTER 3

The Interfacial Region

3.1 INTRODUCTION

Electron transfer at an electrode is a molecular scale event involving the movement of the negatively charged electron between the electrode and a species in solution. The driving force for the transfer of electrons is the gradient in potential created by the application of a potential to the electrode. The gradient in potential is substantial; the local potential difference between the electrode surface and the electrolyte solution can easily be 1 V and this potential change may occur over a fraction of 1 nm, giving a potential gradient $> 10^7 \, \text{V cm}^{-1}$. In this chapter, the convention of discussing the potential distribution in the interfacial region in terms of the Galvani potentials, ϕ, will be followed. These local potentials cannot be measured or controlled in a direct way but it is generally assumed that any change in potential of the electrode *versus* a reference electrode leads to a corresponding change in the potential gradient across the interface, *i.e.*:

$$\Delta E = \Delta \eta = \Delta(\phi_M - \phi_S) \tag{3.1}$$

where ϕ_M and ϕ_S are the Galvani potentials on the metal surface and in the bulk solution, respectively.

The application of a potential to the electrode causes the surface of the electrode to take up a characteristic charge (also dependent on the electrode material and the solution composition). This has electrostatic consequences within the electrolyte. Oppositely charged ions as well as dipoles are attracted to the surface and an 'electrical double layer' is

A First Course in Electrode Processes, 2nd Edition
By Derek Pletcher
© Derek Pletcher 2009
Published by the Royal Society of Chemistry, www.rsc.org

formed. The organization of the ions close to the surface determines the distribution of potential as a function of distance from the surface and hence the driving force for electron transfer. Depending on the potential applied to the electrode, the surface charge can be positive or negative, leading to the attraction of anions or cations, respectively, to the surface. Clearly, there must also be a potential where the surface charge is zero and neither anions nor cations are attracted to the surface; this potential is known as the potential of zero charge, E_{pzc}.

Experiments to study the structure of the interfacial region need to reflect only the changes to surface charge and ion distribution in the interfacial region. Hence, the experiments must be carried out under conditions where there is no transfer of charge across the interface leading to chemical change, either to the chemical composition of the solution or the electrode surface (*i.e.* no Faradaic current is flowing). An electrode where no Faradaic current flows on changing the potential is known as an 'ideally polarized electrode'. Most electrode materials cannot provide these conditions (due to O_2 or H_2 evolution or changes to the chemical composition of the surface with potential). The exception is mercury in some aqueous solutions; over a significant potential range, the currents for surface oxidation/reduction and hydrogen evolution are negligible and the non-Faradaic processes can be examined without significant interference from Faradaic reactions. The mercury/aqueous electrolyte interface is the nearest approach available to an 'ideally polarized electrode'. The assumption is, however, that the concepts developed from studies of this interface are generally applicable to other electrode materials and solutions.

Most commonly, the 'ideally polarized electrode' is studied by measurement of the capacitance, C, defined by the change in charge on the electrode surface, q_M, with change in potential:

$$C = \frac{\mathrm{d}q_M}{\mathrm{d}E} \tag{3.2}$$

The capacitance can be measured in several ways:

1. With an AC capacitance bridge or, now more commonly, in an AC impedance experiment (Chapter 7).
2. From a cyclic voltammogram where, in the absence of Faradaic processes, the observed current density is given by:

$$j = C\nu \tag{3.3}$$

where ν is the potential scan rate.

3. In the case of the liquid metal, mercury, it is conveniently determined from measurements of the surface tension of the interface, γ, as a function of potential. The capacitance is then estimated from:

$$C = -\frac{d^2\gamma}{dE^2} \tag{3.4}$$

This method clearly shows changes at the interface, *e.g.* the adsorption of organics, but has limited precision for the determination of capacitance because of the difficulty in determining the second differential accurately from the experimental data.

This chapter is about the models used to understand the interfacial region and also considers some of the experimental consequences of the existence of a double layer. It needs to be recognized that, in general, the interfacial layer being discussed here is generally <1 nm thick and it should not be confused with the mass transport layers discussed elsewhere in this book, since the thickness of a mass transport layer will typically be microns.

3.2 MODELS OF THE ELECTRICAL DOUBLE LAYER

As is common in science, the complexity of the models used, in this case to describe the interfacial region at an ideally polarized electrode, has increased with time. Figure 3.1 illustrates the models for the case when the potential is negative to the potential of zero charge; also shown are the consequent plots of local potential *versus* the distance perpendicular to the surface. The first model was proposed by Helmholtz (Figure 3.1a), who envisaged that all the charge on the metal is balanced by a monolayer of ions of opposite charge immediately adjacent to the surface. As a result, the potential changes steeply and linearly over the thickness of this monolayer. Outside this layer, the solution will have the composition (and potential) of the bulk with equal concentrations of anions and cations moving randomly through the solution. Somewhat later, it was noted that the ability of ions to move freely through an electrolyte solution at ambient temperatures should be taken into account. In consequence, a second model was suggested (Figure 3.1b). Again the surface charge is balanced by ions in solution of opposite charge but the ions are now free to move in solution rather than bound to the surface; close to the surface there is a large excess of cations over anions and this difference decreases smoothly over a few nanometres

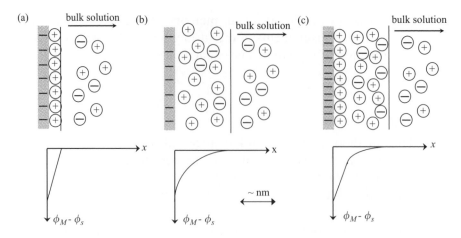

Figure 3.1 Schematic representation of the (*a*) Helmholtz, (*b*) Gouy–Chapman and (*c*) Gouy–Chapman–Stern models for the electrical double layer at an electrode/electrolyte interface. Below each model are the resulting distributions in the potential perpendicular to the surface.

until the composition of the bulk solution is reached. In such a model the potential will drop smoothly across the whole of the layer with a composition different from the bulk, most rapidly at the surface and with a decreasing gradient across this layer. The final model, Figure 3.1(c), combines the two concepts. This model envisages a compact layer of ions at the surface and a diffuse layer where cations again outnumber anions with a gradual change to the bulk solution composition with distance from the surface. Now, a significant fraction of the potential change will occur linearly with distance across the compact layer with the rest dropped as a smooth function of distance across the diffuse layer.

These early models do not take into account several facts. Firstly, the ions, particularly the cations, will be hydrated and the hydration shell increases their size substantially and moves the plane of closest approach, known as the ϕ_2-plane, away from the surface. In fact, it may be necessary to consider two planes of closest approach. The first results when it is the hydrated ions in contact with the surface and is known as the outer Helmholtz plane. The second considers specific adsorption of the ions, most commonly observed with anions because of their greater ability to form chemical bonds with metals; here the hydration shells are partially or fully stripped from the ion and some covalent bonding between the surface and the ion is envisaged. The ion is therefore positioned closer to the surface and the plane of closest approach is known as the inner Helmholtz plane. It should then be noted that, when there is specific adsorption, the potential of zero charge can correspond

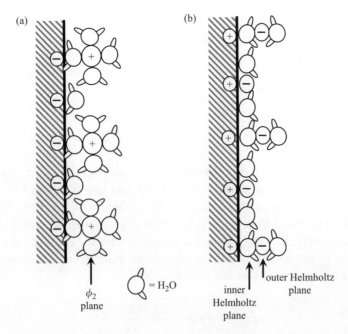

Figure 3.2 Model of the compact double layer: (*a*) negative and (*b*) positive to the point of zero charge.

to the situation where the charges from cations and anions in the double layer just balance. Finally, due to their dipole moments, the water molecules will adsorb on the surface and effectively the solvent and ions will compete for sites on the surface. This changes the structure of the compact layer and influences the distribution of charges.

Figure 3.2(a) and (b) summarizes the upgraded model of the compact double layer negative and positive to the potential of zero charge, respectively (for clarity, the diffuse double layer is not shown). Negative to the potential of zero charge, the surface layer is made up of hydrated cations and water molecules; while the presence of specifically adsorbed anions or cations cannot be ruled out, they are unusual. Positive to the potential of zero charge, both hydrated and specifically adsorbed anions are likely as well as adsorbed water. The presence of the two types of anions leads to splitting of the ϕ_2-plane into the inner and outer Helmholtz planes.

As the compact and diffuse double layers will each have a capacitance, the interfacial region will behave as a pair of capacitances in series and, in these circumstances, the measured capacitance, C, is given by:

$$\frac{1}{C} = \frac{1}{C_{compact}} + \frac{1}{C_{diffuse}} \tag{3.5}$$

where $C_{compact}$ and $C_{diffuse}$ are the capacitances of the compact and diffuse layers, respectively. It can be seen that the measured capacitance is determined by the smallest of these two capacitances. In addition, the potential change between the metal surface and the bulk solution clearly has two contributions identified with the compact and diffuse layers, *i.e.*:

$$\phi_M - \phi_S = (\phi_M - \phi_2) + (\phi_2 - \phi_S) \tag{3.6}$$

A more quantitative description of the compact double layer would first require an exact structure of the layer and this is beyond the ability of the qualitative models to predict. In contrast, a more quantitative description of the diffuse layer is possible if it is assumed that Maxwell–Boltzmann statistics may be used to relate the local concentration of ions at a point in the diffuse layer to the bulk concentration *via* the difference in potentials at the two points, *i.e.*:

$$c_i(x) = c_i \exp - \frac{z_i F[\phi(x) - \phi_s]}{RT} \tag{3.7}$$

The potential is related to charge density, $\rho(x)$. If the ions in solution behave as point charges they are related by the one dimensional Poisson equation:

$$\frac{d^2\phi(x)}{dx^2} = -\frac{\rho(x)}{\varepsilon\varepsilon_o} \tag{3.8}$$

where ε is the dielectric constant of the medium (assumed to be independent of x) and ε_o is the permittivity of free space. The charge density is related to the local concentration by:

$$\rho(x) = \sum_i z_i F c_i(x) \tag{3.9}$$

where for a solution with one electrolyte the series has only two terms, one for the cation and one for the anion. Combining Equations (3.7–3.9) gives:

$$\frac{d^2\phi(x)}{dx^2} = -\sum_i \frac{z_i F c_i}{\varepsilon\varepsilon_o} \exp - \frac{z_i F[\phi(x) - \phi_s]}{RT} \tag{3.10}$$

This differential equation can be solved with the boundary conditions: at the plane of closest approach, $x = x_2$, $\phi = \phi_2$ and away from the surface, $x = \infty$, $\phi = \phi_S$ to give for a 1:1 electrolyte:

$$C_{\text{diffuse}} = \frac{dq_{\text{diffuse}}}{d(\phi_2 - \phi_S)} = F\left(\frac{2\varepsilon\varepsilon_o c}{RT}\right)^{1/2} \cosh\frac{F(\phi_2 - \phi_S)}{2RT} \qquad (3.11)$$

Since at the potential of zero charge, by definition, the surface is uncharged and there is no organization of the charges at the surface, *i.e.* there is no double layer, $\phi_2 = \phi_S$. Hence, Equation (3.11) predicts that a plot of C_{diffuse} *vs* ϕ_2 will be an inverted parabola centred around the potential of zero charge. It also predicts that the diffuse layer capacitance at the potential of zero charge decreases with decreasing concentration of the electrolyte.

Figure 3.3 reports the experimental capacitance for a mercury electrode in solutions of sodium fluoride (1 mM–1 M), selected because fluoride has a very low tendency towards specific adsorption. A feature is the minimum that occurs around the potential of zero charge for the low concentration of electrolyte and it is also clear that the capacitance at the potential of zero charge increases with increasing sodium fluoride ion concentration. These are the characteristics for the diffuse layer capacitance predicted by Equation (3.11) and hence it is reasonable to conclude that this part of the response is associated with the capacitance

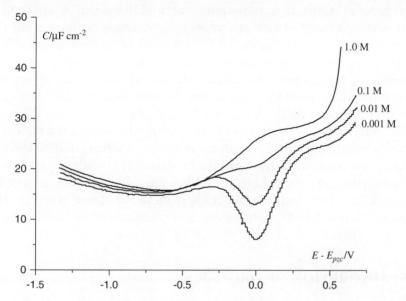

Figure 3.3 Influence of electrolyte concentration on the capacitance as a function of potential for a mercury/aqueous sodium fluoride interface. Data taken from D. C. Grahame, *J. Am. Chem. Soc.*, 1954, **76**, 4819.

of the diffuse double layer. Clearly, away from the immediate vicinity of the potential of zero charge and at high sodium fluoride concentrations the data do not follow the predictions of Equation (3.11) and it is therefore possible to conclude that, under these conditions, the measured capacitance is determined by the capacitance of the compact layer. Equation (3.5) indicates that the measured capacitance is determined by the lowest of the compact and diffuse layer capacitances. We can therefore conclude that the diffuse layer capacitance is low around the potential of zero charge but away from this potential it increases steeply. In contrast, the compact layer capacitance varies relatively little with potential. Some change is seen around the potential of zero charge and this is associated with changes to the orientation of water molecules since there are few ions in the double layer at this point. On the two sides of the potential of zero charge, the capacitance of the compact layer is different and this results from a different ϕ_2 plane with the strongly hydrated sodium ions and the poorly solvated fluoride ions. Also, at quite positive values of the potential, some specific adsorption of fluoride may occur, particularly with concentrated sodium fluoride solutions.

Much stronger increases in capacitance at the potentials positive to the potential of zero charge are seen when the solution contains other anions such as chloride or bromide. Specific adsorption of such halide ions can even occur at relatively negative potentials and this is seen by a negative shift in the potential of zero charge with increasing concentration of the electrolyte (not seen in Figure 3.3 for solutions of sodium fluoride); the point of zero charge then results from a double layer containing balancing amounts of both anion and cation. In general, the extent of specific adsorption at the mercury/aqueous solution interface increases along the series: $F^- < PF_6^- < BF_4^- < ClO_4^- < Cl^- < Br^- < I^-$.

It can be seen that the model of the interfacial region of the mercury/aqueous electrolyte solution with a compact and diffuse double layer is entirely consistent with the experimental data and provides a good basis for understanding its behaviour. Moreover, although the interfacial region with other electrode materials is inevitably more complex, the ideas developed for the mercury electrode continue to be constructive.

3.3 ADSORPTION OF ORGANICS

When neutral organic molecules are present in the aqueous electrolyte they will compete with the ions and water molecules for sites within the compact double layer on the mercury surface. In general, when the organic molecules adsorb on the surface they displace ions, and when they desorb from the surface the ions enter the compact layer; such changes correspond to a significant change in charge distribution at the surface. During a potential scan experiment, such changes to the composition of the compact layer will lead to peaks on a capacitance *vs* potential plot. It is to be expected that the organic compound will adsorb most strongly at potentials around the potential of zero charge where competition from ions for sites in the compact layer is at a minimum. In addition, as the concentration of the organic compound is increased, the capacitance peaks will become larger and the peaks corresponding to adsorption/desorption will move apart as the organic is adsorbed over a broader potential range. Figure 3.4 reports capacitance *vs* potential curves for a mercury drop electrode in 0.1 M KCl, with and without 1-pentanol; it can be seen that the trends expected are, indeed, observed. In particular, with the more concentrated solution of 1-pentanol, well-defined peaks are seen at ~ 0 and -1.2 V *vs* SCE and the alcohol is adsorbed in the potential range between these peaks. In the more dilute solution, the peaks are smaller and occur at ~ -0.3 and ~ -0.9 V; the alcohol is adsorbed to a lesser extent and over a narrower potential range.

A very simple method requiring less instrumentation is to measure the electrocapillary curve, *i.e.* a plot of surface tension for the mercury/ solution interface as a function of potential. A dropping mercury

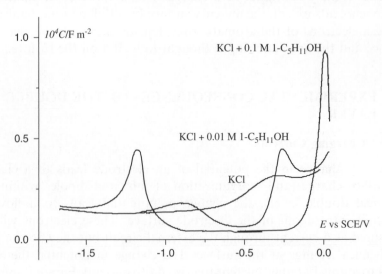

Figure 3.4 Capacitance as a function of potential for a mercury drop electrode in solutions containing 0.1 M KCl, with and without 1-pentanol (see text for details).

electrode consists of a fine glass capillary through which a stream of mercury is forced by applying a head of mercury. The surface tension of the mercury/solution interface can then be determined by placing the dropping mercury electrode into the solution and at each potential (a) determining the height of the mercury head where the interface between the mercury and solution at the end of the capillary is stationary (*i.e.* the electrode no longer forms drops) or (b) slightly less accurately, by measuring the drop time of the electrode with a stopwatch. Figure 3.5 shows electrocapillary curves for solutions of 1-pentanol and phenol in perchloric acid solution. With both compounds, the surface tension is a strong function of their concentrations (indicating adsorption onto the mercury surface) close to the potential of zero charge. In the case of 1-pentanol, however, it can be seen that the surface tension is a function of the alcohol concentration predominantly negative to the maximum in the curve, E_{pzc}, and this potential shifts positively with increasing alcohol concentration. These changes indicate that 1-pentanol is adsorbed negative to the potential of zero charge; the interaction with the mercury surface is stronger with the positive end of the dipole towards the surface. Indeed, the 1-pentanol desorbs at more positive potentials and the surface tension tends to the value for the perchloric acid solution without 1-pentanol. This interpretation of the electrocapillary curves is entirely consistent with the capacitance data of Figure 3.4. In contrast, the data for phenol show a greater sensitivity to phenol concentration positive to the potential of zero charge. This is taken as an indication that phenol adsorbs on the mercury surface by a different mechanism; it is the π-electrons of the aromatic ring that interact with the electrode surface and the aromatic ring is thought to lie flat on the surface.

3.4 EXPERIMENTAL CONSEQUENCES OF THE DOUBLE LAYER

3.4.1 Charging Currents

Since any change in the potential of an electrode leads to a change in surface charge (and reorganization of ions and dipoles within the electrical double layer), the change in potential leads to a flow of electrons into or out of the electrode surface. These electrons will be measured as a current through the external circuit but they do not lead to chemical change at the surface. The change in potential therefore leads to a non-Faradaic current that is additive to any Faradaic current (current corresponding to electrons crossing the electrode/solution interface and bringing about chemical change). The experimental

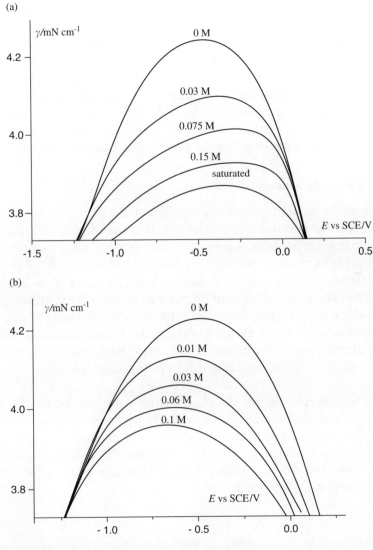

Figure 3.5 Electrocapillary curves for a mercury electrode in 0.1 M HClO$_4$ containing various concentrations of (*a*) 1-pentanol and (*b*) phenol. The concentrations are shown on the plots. Data taken from R. S. Hansen, D. J. Kelsh and D. H. Grantham, *J. Phys. Chem.*, 1963, **67**, 2316.

current density has two contributions:

$$j = j_{\text{Faradaic}} + j_{\text{charging}} \tag{3.12}$$

where the process of changing the electrical double layer is often called 'charging the double layer'.

In many experiments, the objective is the study of the Faradaic process and it is then desirable to minimize the influence of double layer charging. This is part of good experiment design and is discussed further in Chapter 6. Notably, however, once the double layer has reorganized to the structure appropriate to the new potential, no further change will occur. Hence, on changing the potential, the charging current will initially be relatively large but will decay to zero relatively rapidly. As a result, the charging current is normally significant only in experiments with a short timescale or when the potential is scanned rapidly.

3.4.2   Electrode Kinetics

A reasonable model for electron transfer at the electrode surface will assume that the electroactive species is within the compact double layer when the transfer actually occurs. As a consequence, the structure of the double layer might be expected to influence the kinetics of the process. Here, the particular case of a cation being reduced at a potential negative to the potential of zero charge will be considered although the concepts are general. The reduction is likely to occur when the cation is as close as possible to the electrode surface, *i.e.* it is sitting in the plane of closest approach with the centre of the ion at the ϕ_2 plane. With regard to the kinetics of reduction, this will have two consequences:

(a) The concentration of the electroactive cation at the site for electron transfer, the ϕ_2 plane, will not be the same as in the bulk solution just outside the diffuse layer. The concentration of the ion must be adjusted to take into account the change in potential across the diffuse double layer. This can be achieved using Equation (3.7) in the form:

$$c_i(\phi_2) = c_i \exp -\frac{z_i F(\phi_2 - \phi_s)}{RT} \tag{3.13}$$

where $c_i(\phi_2)$ is the concentration of the ion at the ϕ_2 plane. At a potential negative to the point of zero charge, the concentration of a cation at the ϕ_2 plane will be higher than the concentration just outside the double layer.

(b) The driving force for electron transfer will be less than expected. When the reactant is at the ϕ_2 plane, the driving force will only be $(\phi_M - \phi_2)$ not $(\phi_M - \phi_S)$, the potential difference across the whole double layer.

These effects are particularly marked when the ionic strength of the solution is low and the electron transfer is occurring close to the

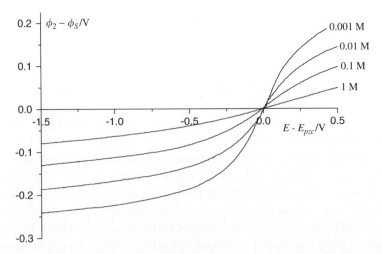

Figure 3.6   Potential difference between the outer Helmholtz plane and the bulk solution potential as a function of electrode potential for a mercury/aqueous sodium fluoride solution interface. Concentrations of NaF are shown on the figure. Data taken from R. Parsons, *Adv. Electrochem. Electrochem. Eng.*, 1961, **1**, 1.

potential of zero charge because the potential drop across the diffuse layer is then a strong function of both the potential and the electrolyte concentration (Figure 3.6).

Because either the reactant or a product of electron transfer is always an ion, appropriate correction to kinetic parameters for the existence of the double layer should be considered, whether the reactant is a cation, anion or neutral species. In reality, such corrections are important only when fundamental interpretation is attempted of (a) rate constants for electron transfer or (b) the slope of a current *versus* potential characteristic (*e.g.* a Tafel plot). Examples would be (a) the comparison of standard rate constants for reactions occurring over a wide potential range and (b) seeking to decide whether the transfer coefficient is a function of potential (Chapter 4). Figure 3.6 shows that the potential drop across the diffuse double layer is a strong function of potential close to the potential of zero charge and this can lead to unusual voltammetric responses. Away from this potential region, the magnitude of the potential drop is larger but it is almost independent of the potential, leading to less obvious effects to voltammetric responses.

3.5   CONCLUDING REMARKS

The discussion of the electrical double layer in this chapter has emphasized an ideally polarized electrode and data are restricted to the mercury/aqueous solution interface. It is, however, evident that the models of the interfacial region fit well to the data of the mercury electrode in aqueous solutions.

A double layer will exist at all electrode/solution interfaces. Because of Faradaic processes, including oxidation and reduction of the surface, the double layer at other metals is more difficult to characterize and/or study. In general, information about the double layer (and, for example, the potential of zero charge) is deduced less directly from experimental data and the method is often specific to the system under study. In consequence, relatively little data exist about the double layer structure for electrodes other than mercury. Even so, a double layer will exist at all electrode/solution interfaces and it will influence the behaviour of the electrode and data from electrochemical experiments. The ideas developed with the ideally polarized electrode can, at least qualitatively, be applied to such electrodes.

It should be stressed that all the conclusions about the composition and structure of the double layer are based on purely electrostatic arguments. A corollary is that when electrochemical experiments are carried out with a large excess of electrolyte in solution it is the ions of this electrolyte that control the structure of the double layer formed. In addition, when chemistry is occurring at the surface, for example, when covalent bonds between an adsorbate and the surface are formed, the model of the interfacial region will need substantial modification.

Further Reading

1. P. Delahay, *Double Layer and Electrode Kinetics*, Interscience, New York, 1965.
2. M. Sparnaay, *Electrical Double Layer*, Pergamon Press, Oxford, 1972.
3. R. Parsons, *Modern Aspects Electrochem.*, 1954, **1**, 103.
4. R. Parsons, *Advances in Electrochemistry, Volume 1*, ed. P. Delahay, Interscience, New York, 1961, p. 1.
5. Southampton Electrochemistry Group, *Instrumental Methods in Electrochemistry*, Ellis Horwood, Chichester, republished 2001.
6. J. O'M. Bockris and A. K. N. Reddy, *Modern Electrochemistry, Volume 2A – Fundamentals of Electrodics*, Kluwer Academic, New York, 2001.

CHAPTER 4

A Further Look at Electron Transfer

4.1 INTRODUCTION

The thermodynamics and kinetics of the electrode reaction:

$$O + ne^- \rightleftharpoons R \tag{4.1}$$

were introduced in Section 1.3. It was shown that, at equilibrium (*i.e.* no current flow), the potential of an inert electrode in a solution containing O and R (as well as an excess of inert electrolyte) depends on the formal potential of the couple O/R and the concentration of the two species in solution [the Nernst equation, Equation (1.29)]:

$$E_e = E_e^\circ + \frac{2.3RT}{nF} \log \frac{c_O}{c_R} \tag{4.2}$$

Away from the equilibrium potential, the kinetics of electron transfer were described by Equations (1.43–1.49), most importantly the Butler–Volmer equation at low overpotentials:

$$j = j_o \left(\exp \frac{\alpha nF\eta}{RT} - \exp -\frac{(1-\alpha)nF\eta}{RT} \right) \tag{4.3}$$

and the Tafel equation at higher overpotentials (written for an oxidation):

$$\log j = \log j_o + \frac{\alpha nF}{2.3RT} \eta \tag{4.4}$$

A First Course in Electrode Processes, 2nd Edition
By Derek Pletcher
Published by the Royal Society of Chemistry, www.rsc.org

It was concluded that the experimental current density, j, depended on the applied overpotential, η, the exchange current density, j_o (or the standard rate constant) and the transfer coefficient, α. The transfer coefficient was considered to be a constant, usually equal to 0.5. While they are usually adequate for the experimental electrochemist, these equations are, at best, semi-empirical. They certainly contain assumptions, *e.g.* the form of the potential dependence of the rate constants, and also give little understanding of the process of electron transfer on a molecular level. This chapter, therefore, examines the electron transfer in more detail.

Throughout this chapter it will be assumed that the current density is well below the value where there is significant depletion of reactant close to the electrode surface, *i.e.* mass transport limitations need not be taken into account.

4.2 FORMAL POTENTIAL

It is important to recognize that the formal potential for the couple is determined entirely by the chemistry of O and R. Figure 4.1 shows a simple thermodynamic cycle for the reduction of a solvated species. It can be seen that the Gibbs free energy for the reduction O/R is determined by (a) the relative values for the Gibbs free energies of solvation of O and R and (b) the electron affinity of O determined by the energy of the lowest unfilled orbital on the electroactive species in the gas phase. Such a cycle also leads to an understanding of the influence of complexing agents. Addition of a ligand to the electrolyte will change (usually increasing) the Gibbs free energies associated with the conversion of the metal complex in solution into an uncomplexed metal ion in the gas phase. Since, in general, the ligand will interact more strongly with

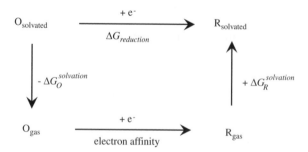

Figure 4.1   Simple thermodynamic cycle for the reduction of O to R at an electrode surface.

the higher oxidation state of the metal, adding the ligand will make the reduction more difficult and shift the formal potential more negative.

The Gibbs free energy for the reduction (or oxidation) under standard conditions, and with a solution where $c_O = c_R$, is related to the formal potential by the equation:

$$\Delta G^0 = -nFE_e^{\circ} \qquad (4.5)$$

but it should be noted that the formal potential is measured *versus a particular reference electrode*, for example the standard hydrogen electrode where the reaction is:

$$2H^+ + 2e^- \rightleftharpoons H_2 \qquad (4.6)$$

so that the Gibbs free energy being discussed is that for the reaction:

$$O + H_2 \rightleftharpoons R + 2H^+ \qquad (4.7)$$

including the chemistry at the reference electrode used.

4.3 KINETICS OF ELECTRON TRANSFER

4.3.1 Setting the Scene

When electron transfer is occurring between a metal electrode and a species in solution, the electron sees the electrode/solution interface as a total discontinuity in its environment. Within the metal, the electrons within the conduction band are delocalized and move freely about the metal lattice. Essentially, there is a large number of closely spaced energy levels around the Fermi level, both filled levels immediately below the Fermi level and empty ones immediately above it, and, in consequence, adding or removing an electron should not be a difficult process. In contrast, on the solution side, the electrons reside within localized orbitals on specific ions or molecules and the highest filled orbital or lowest unfilled orbital can have an energy close to or well away from the energy of the Fermi level within the metal. It also has to be recognized that, at ambient temperatures, the solution environment is constantly fluctuating – all species are free to move, the movement of ions (the reactant if charged and ions of the electrolyte) leading to changes in the electrostatic environment, bonds within the electroactive species are vibrating and rotating and solvent molecules within the solvation shell can reorientate. As a result, the electronic energy levels

Table 4.1 Typical timescales of fundamental steps involving electroactive species in solution during an electron-transfer process.

Step	Typical timescale/s
Electron transfer	10^{-16}
Alteration to bond length	10^{-14}
Reorientation of solvent molecule	10^{-11}
Reorientation of ionic atmosphere	10^{-8}
Bond cleavage or structural change	$> 10^{-8}$

within the electroactive species are time variant and changing continuously; it is possible only to view each electronic energy level as a distribution function (usually Gaussian) around the 'most probable' energy. It is helpful to recognize the typical timescales of the events occurring during the electron-transfer event; they are set out in Table 4.1.

In view of the electronic environments on the two sides of the interface, the focus of attention must be on the solution side of the interface. Moreover, with the quite different timescales of the fundamental steps in the chemical change from reactant to product, the electron-transfer reaction can be envisaged as a three-step process: (a) reorganization of the environment of the electroactive species to an intermediate structure, (b) electron transfer and (c) relaxation of the intermediate structure to that of the stable product of electron transfer. Electron transfer, being by far the fastest event, will occur while the solution side of the interface appears 'frozen'. Figure 4.2 illustrates this mechanism. Electron transfer is subject to the Franck–Condon principle. This states that the probability of electron transfer is highest when the initial and final electronic states involved have the same energy. Hence, in the model, the intermediate state is the reactant with its environment reorganized so that the reactant and product both have an electronic state with the same energy. Electron transfer would then be followed by a reorganization of the product environment to its equilibrium state. Clearly, the expectation must be that the intermediate structure will be somewhere between the equilibrium states for the reactant and product.

4.3.2 Absolute Rate Theory

In absolute rate theory, the rate constant of a chemical reaction is interpreted in terms of an energy of activation. It is envisaged that the reaction passes through a 'transition state' or 'activated complex' with an intermediate structure between reactants and products. The formation of this transition state requires an input of energy, the energy

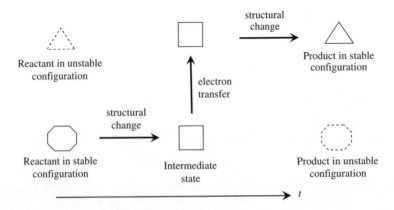

Figure 4.2 Model for an electron-transfer reaction.

of activation. Once the transition state is formed, however, conversion into product as well as return to the reactants are favourable processes. For a heterogeneous chemical reaction, the rate constant is related to the energy of activation, $\Delta G^{\ddagger}$, by:

$$k = KZ \exp - \frac{\Delta G^{\ddagger}}{RT} \qquad (4.8)$$

where K is known as the transmission coefficient and $Z = \delta k_{B} T / h$ with δ a reaction length of the order of a molecular diameter. It is often more convenient to interpret KZ in terms of an upper limit to the rate constant usually associated with the frequency of vibration of the activated complex; this sets the value between 1 and $10 \, \mathrm{cm \, s^{-1}}$.

For a simple electron-transfer reaction such as Equation (4.1), the concepts of absolute rate theory are illustrated by the reaction coordinate curves of Figure 4.3, drawn for the case where $c_{O} = c_{R}$. At the equilibrium potential, Figure 4.3(a), it should be noted that (a) the energy levels of the reactant and product are identical since the Gibbs free energy change for the reaction is zero and (b) since the rate of the forward and back reactions are equal, the energies of activation for these reactions are the same. At other potentials, the potential energy surfaces for both reactant and product will be shifted. At a potential negative to the equilibrium potential (Figure 4.3b) the changes must lead to a negative Gibbs free energy for the reduction step and the energy of activation for reduction must be less than that for the oxidation. At potentials positive to the equilibrium potential (Figure 4.3c), in contrast, there must be a negative Gibbs free energy for the oxidation step and the energy of activation for oxidation must be less than that for the reduction.

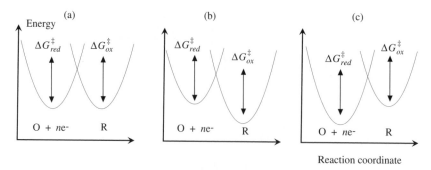

Figure 4.3 Energy curves for the electrode reaction $O + ne^- \rightarrow R$ for $c_O = c_R$: (*a*) at equilibrium at the formal potential – no net current density, (*b*) negative to E_e – net reduction occurs and (*c*) positive to E_e – net oxidation occurs.

We have noted earlier that a change in the potential applied to the electrode, ΔE, leads to a change in the local potential difference across the electrode/solution interface, $\Delta(\phi_M - \phi_S)$, and it is the resulting potential field that drives the movement of the electron across the interface. In considering further the influence of potential on the rate of electron transfer it is convenient to consider that the change in potential influences only the solution potential, ϕ_S; that is, making the potential of the electrode more negative by ΔE has the effect of making the solution potential, ϕ_S, more positive by ΔE. This procedure is consistent with noting that it is only the local *potential difference* that is critical and reasonable with our models of the two sides of the interface. It is also equivalent to earthing the electrode.

Two specific reductions will now be considered. The first is the reduction of a M^+ ion at an electrode fabricated from the metal M or having a coating of M:

$$M^+ + e^- \longrightarrow M \tag{4.9}$$

The potential energy surfaces for two potentials are sketched in Figure 4.4, assuming that the surfaces are parabolic. On the right-hand side, the surfaces are shown for the potential, E. When the potential is made more negative to $E - \Delta E$, this is equivalent to increasing the solution potential by $+\Delta E$ and thereby destabilizing the M^+ cation by an energy of $F\Delta E$. As a result, the potential energy surface for M^+ is shifted up by $F\Delta E$. The potential energy surface for the product, as a metallic deposit on the electrode, is unaffected by the change in solution potential.

From Figure 4.4, the change in potential clearly leads to a decrease in the activation energy. Measurement with a ruler will show that in fact the decrease in activation energy is $1/2F\Delta E$, *i.e.* only half of the change

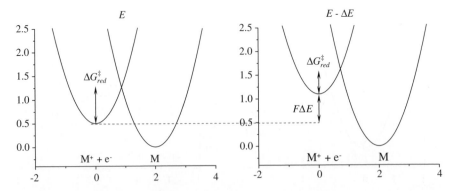

Figure 4.4 Energy curves for the reduction of M^+ to M (metal of the electrode surface) at the potentials E and $E - \Delta E$. The scales on the axes are in arbitrary units and are shown only to guide the reader.

in potential is successful in changing the rate of reduction. It can also be deduced that:

$$\frac{d \ln j_{red}}{dE} = -\frac{1}{2} \frac{F}{RT} \qquad (4.10)$$

and this is equivalent to the differential of the Tafel equation, Equation (1.47) if $\alpha = 1/2$. This relationship between the change in potential and the change in activation energy is one interpretation of the transfer coefficient, α, in the Butler–Volmer and Tafel equations – only a fraction, α, of the change in potential is successful in changing the energy of activation for the electron-transfer reaction.

A similar diagram can be drawn for the reaction:

$$Fe^{3+} + e^- \longrightarrow Fe^{2+} \qquad (4.11)$$

Using the same argument, making the electrode potential more negative effectively destabilizes cationic species in the solution. But the ferric and ferrous ions will be destabilized by different amounts due to their different charges, $3F\Delta E$ and $2F\Delta E$, respectively. This is shown in Figure 4.5 and again it can be seen that the energy of activation for reduction is decreased at the more negative potential. Careful measurement will show that the decrease is again $1/2F\Delta E$ for a change in potential of ΔE.

The discussion so far assumes that the potential energy surfaces are identical parabolas, equivalent to assuming that the two systems may be treated as simple harmonic oscillators with the same force constants. It can also be seen that the conclusions are valid only over a range of ΔE.

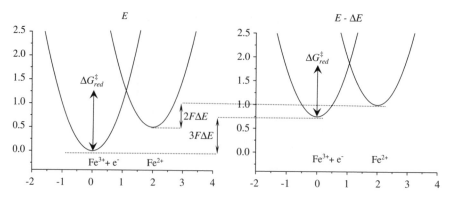

Figure 4.5   Energy curves for the reduction of Fe^{3+} to Fe^{2+} at the potentials E and $E - \Delta E$. The scales on the axes are in arbitrary units and are shown only to guide the reader.

Larger changes in potential will lead to intersection of the surfaces at points where the curves have very different slopes and then the transfer coefficient will become potential dependent. For the case of a solution where $c_O = c_R$, it can be shown using straightforward geometry that a Tafel equation is obtained only if the transfer coefficient is given by the expression:

$$\alpha = \frac{1}{2} + \frac{nF\Delta\eta}{16^\circ\Delta G^{\ddagger}} \qquad (4.12)$$

where $^\circ\Delta G^{\ddagger}$ is the equal activation energy for reduction and oxidation at the equilibrium potential. The value of the transfer coefficient appears to vary linearly with potential and a Tafel plot should be nonlinear and even pass through a maximum in extreme conditions. The extent of the nonlinearity depends on the Gibbs free energy of activation at the equilibrium potential and the overpotential. Since a low value of $^\circ\Delta G^{\ddagger}$ equates to a fast reaction, the variation of the transfer coefficient with potential will be most marked for very fast reactions. It will also be more pronounced at large overpotential. The study of fast reactions at high overpotentials is very difficult because of interference from mass transfer control and experimental limitations as well as the requirement to correct the data for double layer effects. This has limited our ability to confirm convincingly by experiment the potential variation of the transfer coefficient. Certainly, for most experimental studies it is sufficient to consider the transfer coefficient as a constant, independent of potential.

4.3.3 Fluctuating Energy Level Model for Electron Transfer

In this model, emphasis is placed on a consideration of the fluctuations of the energy levels within the reactant and product of electron transfer due to thermally induced motion of their surroundings, the solvent molecules and/or ligands as well as vibrations/rotations of bonds within the species. The theory applies to electrode reactions that involve only electron transfer and are therefore akin to an outer sphere mechanism for electron transfer in homogeneous solution; no bonds are broken or formed and the inner coordination shell remains unchanged during the oxidation/reduction.

As noted in the discussion of Figure 4.2, the transition state for an electrode reaction is envisaged as a non-equilibrium species with an arrangement of surrounding solvent/ligands intermediate between the equilibrium states of the reactant and product in the electrolyte solution. The transition state must have an electronic energy level exactly matching an energy level within the electrode metal. Electron transfer occurs only when this particular structure is reached through thermally induced vibrations and rotations within the reactant and motion of solvent/ions in the surroundings. As also noted earlier, see Table 4.1, the actual transfer of the electron is rapid compared with the rearrangement of the surroundings to the reactant and the theoretical description of electron transfer is advanced by recognizing that, since they have quite different timescales, the reorganization leading to the transition state and the electron-transfer event may be treated separately.

The first quantitative descriptions of electron transfer using quantum mechanics were developed by Marcus and Levich with Dogonadze, who recognized that, because of the different timescales, the situation can be simplified using the Born–Oppenheimer approximation to separate the Hamiltonians for the electrons and the nuclei. Indeed, the slow re-organization of the reactant and product environments (solvent, ligands, other ions, *etc.*) is adequately handled by semi-classical models. In contrast, the electrons must be treated using quantum mechanics and, since they are small particles, the treatment must include tunnelling of the electrons through the potential energy barrier separating reactant and product.

Marcus used a 'hard sphere in a dielectric continuum' model previously employed to estimate the free energy of solvation of ions but modified it to include ligands interacting with the electron-transfer site. The aim was to estimate the reorganization energy, λ, associated with formation of the activated complex. The core assumption is that small fluctuations in the surroundings to the electron-transfer site can lead to

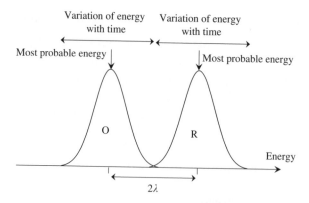

Figure 4.6 Temporal probability distribution for electron states of the species O and R. The variations arise from changes to the environment, *e.g.* vibration of bonds, movement of surrounding solvent molecules and ions in the double layer.

large variations in the electronic energy levels. The energy levels of all the reactant and product molecules/ions in the system are constantly fluctuating (*i.e.* varying rapidly with time) about a most probable energy level as the environment of the electron-transfer centre changes due to vibration and rotation of bonds within the molecule/ion or movement of solvent molecules/ions surrounding the reactant. The distribution of energies about the most probable values for the oxidized and reduced species are represented as bell-shaped Gaussian curves in Figure 4.6. The difference in charge on the oxidized and reduced species has led to a splitting of the most probable energy levels by 2λ. The intersection of the two curves gives the redox potential for the solution (*i.e.* the equilibrium potential calculated from the Nernst equation).

Figure 4.7 shows how these energy levels are used to understand the kinetics of electron transfer. As discussed in Chapter 2, within the metal there is a large number of closely spaced, delocalized energy levels above and below the Fermi level (the highest filled level). Within the solution, the localized energy levels on the reactant and product (*e.g.* for an octahedral transition metal complex the level of interest will be an e_g or t_{2g} level from the splitting of the d orbitals) are fluctuating in energy as shown in Figure 4.6. The situation at the equilibrium potential is shown in Figure 4.7(a) and (b); no net electron transfer will occur since the Fermi level in the metal and the redox level in solution are the same. The exchange current density (the amount of balancing oxidation and reduction) is determined by the extent of the overlap of the distribution functions for the oxidized and reduced species (see enlarged picture, Figure 4.7b). Clearly, as the reorganization energy increases, the two

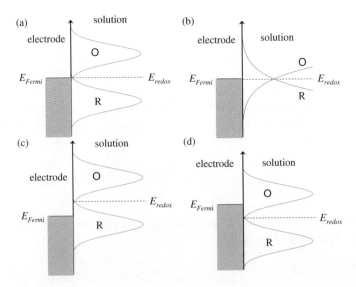

Figure 4.7 Energy level diagrams for: (*a*) and (*b*) the equilibrium potential, (*c*) a positive overpotential and (*d*) a negative overpotential. Here E_{Fermi} and E_{redox} are, respectively, the energies of the Fermi level and a solution energy level equivalent to the equilibrium potential of the couple O/R in solution.

distributions will intersect at a lower probability for the transition state to be formed. Figure 4.7(c) and (d) shows the situation at positive and negative overpotentials, respectively. To be consistent with earlier discussions, the diagrams are sketched assuming that a change in potential leads only to changes in the energy levels for the solution soluble species; it would, however, be equally appropriate to shift the Fermi level in the metal and to assume that the solution levels are unchanged – it is only the potential difference across the interface that is important. At these potentials, electrons will be transferred across the interface until the Fermi and redox levels are again equal (a new equilibrium with changed solution concentrations of O and R has been established). It should be stressed that electron transfer occurs only when there are filled and unfilled energy levels of equal energy on the two sides of the interface and this can occur only because the energy level of the solution species are fluctuating with time and the metal has many unfilled energy levels above the Fermi level.

Quantitatively, the theory allows a calculation of the reorganization energy, λ, and the exact form of the distribution functions and hence the degree of overlap between the distribution functions for the oxidized and reduced species. The calculations confirm the assumption that small changes to the environment of the electron-transfer centre can lead to large variation in energy levels; the reorganization energies calculated were of the order of 1 eV and this led to mean square fluctuations of >0.3 eV compared to $k_B T \approx 0.025$ eV at room temperature. The theory again leads to expressions for the relationships between current density and overpotential that are similar to the Butler–Volmer and Tafel equations but with a transfer coefficient that varies with potential:

$$\alpha = \frac{1}{2} + \frac{nF\Delta\eta}{4\lambda} \qquad (4.13)$$

It can be seen that this expression is the same as that derived from absolute rate theory provided:

$$\frac{\lambda}{4} = {}^\circ\Delta G^{\ddagger} \qquad (4.14)$$

The results of the fluctuating energy level model are therefore formally equivalent to the expressions obtained from absolute rate theory with parabolic energy curves with identical force constants. It can be seen that deviation of the transfer coefficient from a constant value (0.5) is going to be observed for systems with a small reorganization energy (*i.e.* a fast electron-transfer reaction) and at high overpotential, where experimental study is most difficult.

All the above discussion focuses on the reorganization of the surroundings to the electron-transfer centre and this is clearly seen as the slow step in the overall process. The actual electron transfer between the states in the electrode and the solution phase with equal energy occurs by electron tunnelling. The probability of electron tunnelling is usually discussed in terms of a transmission coefficient, $\kappa_{\text{tunnelling}}$, given by:

$$\kappa_{\text{tunnelling}} = \kappa^\circ_{\text{tunnelling}} \exp -\beta x \qquad (4.15)$$

where x is the distance between the electrode and redox centre in solution, β is a constant determined by the height of the energy barrier and the nature of the medium between the states and $\kappa^\circ_{\text{tunnelling}}$ is a proportionality constant. Clearly, the likelihood of tunnelling depends strongly on the separation of the metal surface and the redox centre in the solution (see the next section). The value of $\kappa_{\text{tunnelling}}$ tends to one when x is small so that the coupling between the reactant and electrode is strong when the reaction is said to be adiabatic. Figure 4.8 shows the

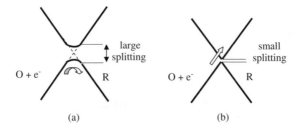

Figure 4.8 Potential energy surfaces close to the intersection of the surfaces for O and R for a strong (*a*) and a weak (*b*) interaction between the metal electrode and redox centre in solution.

potential energy surface close to the point of intersection of the surfaces for reactant and product (the transition state). Strong interaction leads to a splitting of the energy levels for the transition state (into a lower and excited states) that is large compared to kT at room temperature and hence a continuous and lower surface (curve) exists between O and R; tunnelling is likely. Weak interaction leads to a splitting less than kT and it is likely that the system remains within the surface for $O + e^-$; there is a lower probability that the system crosses to the surface for R – tunnelling is less likely.

Most importantly, we can again conclude that in most experimental studies it will be sufficient to consider the transfer coefficient to be a constant; the Butler–Volmer, Tafel and related equations, Equations (1.43–1.49), are generally sufficient for our needs to quantify the kinetics of electron transfer when studying practical electrode reactions.

4.4 SOME EXPERIMENTAL RESULTS

It is, of course, interesting to compare the predictions of theory with experimental observations. While there is an extensive literature on the study of electron-transfer kinetics at electrodes, quantitative comparisons are difficult both because of experimental scatter in the kinetic parameters reported (due to experimental limitations on the techniques, see Chapters 6 and 7, as well as a lack of rigour in controlling/reporting the conditions) and a lack of structural information about many couples in the solutions used for electrochemistry. There is, however, no doubt that the experimental trends agree with the predictions of theory.

In fact, the number of reactions that behave as simple electron-transfer reactions without complications is limited. In aqueous solution, there are several transition metal complexes where both oxidized and reduced forms have octahedral symmetry with similar metal–ligand bond lengths. These include:

$$Fe(CN)_6{}^{3-} + e^- \rightleftharpoons Fe(CN)_6{}^{4-} \tag{4.16}$$

$$IrCl_6{}^{3-} + e^- \rightleftharpoons IrCl_6{}^{4-} \tag{4.17}$$

$$Ru(NH_3)_6{}^{3+} + e^- \rightleftharpoons Ru(NH_3)_6{}^{3+} \tag{4.18}$$

$$MnO_4^- + e^- \rightleftharpoons MnO_4{}^{2-} \tag{4.19}$$

(the latter only in strongly alkaline solution). Consistent with the predictions of theory, all are found to be rapid electron-transfer reactions with rate constants of $k_s > 10^{-2}\,cm\,s^{-1}$. Perhaps surprisingly, hydrated transition metal complexes seldom show such rapid electron-transfer kinetics; the kinetics for the hydrated couples decrease in the order:

$$Eu(III)/Eu(II) > V(III)/V(II) > Ce(IV)/Ce(III)$$
$$\approx Fe(III)/Fe(II) > Cr(III)/Cr(II)$$

with the standard rate constants falling from 10^{-2} to $10^{-5}\,cm\,s^{-1}$. An example of a rapid organometallic system is the oxidation of ferrocene:

$$Cp_2Fe - e^- \rightleftharpoons Cp_2Fe^+ \tag{4.20}$$

where the standard rate constant is $\sim 0.1\,cm\,s^{-1}$. The two cyclopenta-dienyl rings enclose the iron centre tightly so that the ferrocene and ferrocinium ion are both poorly solvated in most solvents and for this reason the equilibrium potential of this couple is often taken to be independent of the solvent. Amongst organic systems, the limitation is usually the chemical instability of the product of electron transfer. An exception is the $1e^-$ reduction of polycyclic aromatic compounds such as anthracene in aprotic solvents such as acetonitrile or dimethylformamide. The electron is inserted into the π-system and the anion radicals are stable. Moreover, the neutral molecule and the anion radical are both flat species so there is little change in structure during the reaction:

$$Ar + e^- \rightleftharpoons Ar^{-\cdot} \tag{4.21}$$

These are the fastest electron-transfer reactions known, with standard rate constants $> 1\,cm\,s^{-1}$.

Another prediction of theory is that there is a relationship between the standard rate constant for heterogeneous electron transfer, k_s, for a couple and the rate constant, k_{homo}, for the homogeneous isotopic exchange reaction between the oxidized and reduced species. For example, there is a relationship between the standard rate constant for the Fe^{3+}/Fe^{2+} couple and the rate constant for the reaction:

$$Fe^{3+} + {}^*Fe^{2+} \rightleftharpoons Fe^{2+} + {}^*Fe^{3+} \tag{4.22}$$

that can be studied by measurement of radioactivity after separation. The relationship derived from Marcus theory is:

$$\frac{k_s}{Z} = \left(\frac{k_{homo}}{Z'}\right)^{1/2} \tag{4.23}$$

where Z and Z' are 'collision numbers' (effectively the fastest possible values for the corresponding rate constants) for the heterogeneous and homogeneous processes, respectively. Although the data available are limited, there is reasonable agreement between the theory and experiment, building confidence in the theory. Both the electrode and homogeneous electron transfers tend to be rapid.

Much slower are couples where electron transfer occurs *via* an inner-sphere mechanism, *i.e.* the reaction requires substitution of a ligand in the inner coordination sphere or substantial steric change (*e.g.* a change from octahedral to tetrahedral geometry). Both heterogeneous and homogeneous electron transfer are then retarded substantially.

4.5 ELECTRON TRANSFER WITH BIOMOLECULES

The electrochemistry of enzymes and other biomolecules is attracting much present interest. Electron transfer at interfaces is essential to life, having a central role in respiration and photosynthesis, and electrochemical techniques can be used in their study. In addition, applications of the electrochemistry of biomolecules are envisaged in biosensors, bio-fuel cells, *etc.*

Electron-transfer enzymes are typically large macromolecules with molecular weights up to 10^6 daltons. They are sophisticated copolymers of amino acids, commonly with one or more transition metal centres (usually, Fe, Co, Cu, Mn or Mo); the superstructure around the metal centres is designed by nature to control the specificity of the chemistry at the active centre. They may also contain cofactors (essentially mediators for electron transfer to the metal centres) within the structure. The chemistry at the metal centre and cofactor may be controlled through the electronic environment, steric factors, hydrophobicity and strain in the metal–ligand bonds and commonly the appropriate environment requires that the metal centre and cofactor are within a cleft and tucked well away from the surface of the macromolecular structure. Figure 4.9 illustrates the structure of such enzymes with the example glucose oxidase. Glucose oxidase is a homodimeric molecule with a hydrodynamic diameter of 17.2 nm and it contains two iron moieties and two flavin adenine dinucleotide (FAD) cofactor centres; its role in nature is to catalyse the oxidation of glucose to gluconolactone by oxygen. Electron transfer occurs initially to the FAD moieties (shown in black) that are hidden within a shell consisting of amino acid chains and carbohydrate substituents (shown in grey) with a complex structure. Clearly, electron transfer to/from such large molecules presents additional factors not considered above. The orientation of the macromolecule to the electrode surface will determine the distance over which the electrons must tunnel and hence determine the standard rate constant for electron transfer (Figure 4.10). The standard rate constants are estimated using Equation (4.15).

Not surprisingly, direct electrochemical oxidation/reduction of enzymes can be a very slow process. One approach to overcoming this problem is to use a mediator. The mediator is a species that undergoes rapid electron transfer at the electrode surface (preferably at a potential close to the equilibrium potential of the redox enzyme) and can then

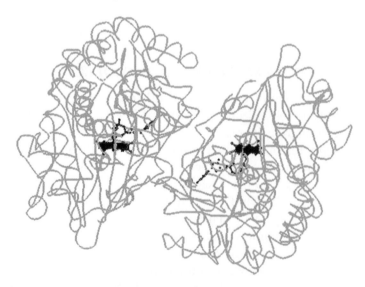

Figure 4.9   Structure of glucose oxidase. The flavin adenine dinucleotide centres involved in electron transfer are shown in black while the surrounding superstructure of amino acids and carbohydrates is shown in grey.

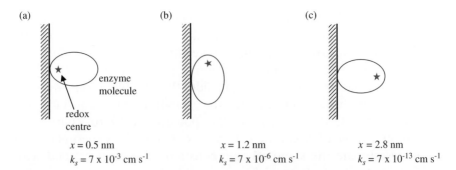

(a) enzyme
molecule

redox
centre

$x = 0.5$ nm
$k_s = 7 \times 10^{-3}$ cm s^{-1}

(b)

$x = 1.2$ nm
$k_s = 7 \times 10^{-6}$ cm s^{-1}

(c)

$x = 2.8$ nm
$k_s = 7 \times 10^{-13}$ cm s^{-1}

Figure 4.10 Influence of molecular orientation of an enzyme molecule at an electrode surface on the distance for electron tunnelling and hence on the standard rate constant for oxidation/reduction.

diffuse to the transition metal centre of the enzyme to return the redox enzyme to its active oxidation state (in nature, oxygen often substitutes for the electrode in the chemistry of oxidases). Many mediators have been reported in the literature, one example being a water-soluble ferrocene.

It is also necessary to recognize that enzymes, being large macromolecules, will have very low values for their diffusion coefficients and hence give very low limiting current densities for their direct electrode reactions. The use of a mediator also helps this limitation but it is much more common to design a modified electrode where the enzyme is a component of an electrode coating; the enzyme can be covalently bound to or adsorbed on the surface. Including the enzyme in an electrode coating also minimizes the amount of this expensive material used in the experiment. It is, however, necessary to check that the stability of the enzyme and the specificity and kinetics of its chemistry are maintained in this unnatural environment.

4.6 CORRECTION OF KINETIC PARAMETERS FOR DOUBLE LAYER EFFECTS

In Chapter 3, the structure of the double layer formed at the interface between the electrode and the electrolyte solution was discussed. It was concluded in Section 3.4.2 that electron transfer was most likely to occur when the reactant was at the plane of closest approach to the surface, the ϕ_2 plane, and this would have two consequences for the rate of electron transfer:

1. The concentration of the electroactive cation at the site for electron transfer, the ϕ_2-plane, will not be the same as in the bulk solution

just outside the diffuse layer. The concentration of the ion must be adjusted to take into account the change in potential across the diffuse double layer. At a potential negative to the potential of zero charge, the concentration of a cation at the ϕ_2 plane will be higher than the concentration just outside the double layer and the experiment will overestimate the standard rate constant for the couple O/R. Positive to the potential of zero charge the concentration of a cation will be lower and the experiment will underestimate the standard rate constant. The reverse trend with potential will be observed for anions.

2. The driving force for electron transfer will be less than expected. When the reactant is at the ϕ_2 plane, the driving force will be only $(\phi_M - \phi_2)$ not $(\phi_M - \phi_S)$, the potential difference across the whole double layer.

In concept, it is not difficult to modify the Butler–Volmer and Tafel equations to take into account these two factors but, to avoid the algebra, only the results are reported here. The true exchange current density is related to the experimental exchange current density by the equation:

$$(j_o)_{exp} = (j_o)_{true} \exp \frac{(\alpha n - z_R)F(\phi_2 - \phi_s)_e}{RT} \tag{4.24}$$

where $(\phi_2 - \phi_s)_e$ is the potential drop across the diffuse double layer at the equilibrium potential for the couple O/R. A similar expression can be written for the standard rate constant:

$$(k_s)_{exp} = (k_s)_{true} \exp \frac{(\alpha n - z_R)F(\phi_2 - \phi_s)_e}{RT} \tag{4.25}$$

In most experimental studies, making these corrections is unnecessary and only adds inconvenience to the interpretation and use of the kinetic parameters; for example, reconstructing the j vs E characteristic would require reversing the double layer corrections. In contrast, correcting the data for the consequences of the existence of the double layer becomes important when seeking to understand the kinetic data at a more fundamental level. To give three examples of where the correction becomes essential:

1. Seeking to understand on a molecular level, the differences in the rate parameters for a series of couples when the equilibrium potentials for the couples occur at different potentials with respect

to the potential of zero charge. The correction factor to the standard rate constants for a series of couples could easily vary between 1 and 100, depending on the values of the equilibrium potentials.

2. Studying the role of the solvent in influencing the rate parameters for a couple. In such studies, the $(\phi_2 - \phi_s)$ *vs* E curves will be required for each of the solvents. In addition, consideration needs to be given to the magnitude of the liquid junction potentials.
3. Investigating whether the transfer coefficient is a function of potential.

Clearly, to make the double layer correction, data are required to construct the variation of $(\phi_2 - \phi_s)$ with E for the electrode/electrolyte combination being used for the study. Such plots were shown in Figure 3.6 for a mercury electrode in aqueous sodium fluoride solutions. It is apparent that the magnitude of $(\phi_2 - \phi_s)$ depends on the concentration of the electrolyte and the potential but can be between 0 and 250 mV. Close to the potential of zero charge, $(\phi_2 - \phi_s)$ varies steeply with potential, and becomes almost independent of potential away from the potential of zero charge. Unfortunately, similar data are not available and are more difficult to determine for other electrode materials and electrolytes and this limits our ability to make reliable double layer corrections in most systems.

Further Reading

1. A. J. Bard and L. R. Faulkner, *Electrochemical Methods – Fundamentals and Applications*, John Wiley & Sons, New York, 2001, ch. 3.
2. Southampton Electrochemistry Group, *Instrumental Methods in Electrochemistry*, Ellis Horwood, Chichester, republished 2001.
3. J. Albery, *Electrode Kinetics*, Clarendon Press, Oxford, 1975.
4. M. J. Weaver, in *Chemical Kinetics, Volume 27*, ed. R. G. Compton, Elsevier, Amsterdam, 1987, p. 1.
5. R. A. Marcus, *J. Phys. Chem.*, 1956, **24**, 966 and *Electrochim. Acta*, 1968, **13**, 995.
6. V. G. Levich, *Advances in Electrochemistry and Electrochemical Engineering, Volume 5*, ed. C. W. Tobias, Interscience, New York, 1967, p. 249.
7. B. Case, R. R. Dogonadze and J. M. Hale, in *Reactions of Molecules at Electrodes*, ed. N. S. Hush, Wiley-Interscience, London, 1971.

8. S. W. Feldberg, M. D. Newton and J. F. Smalley, in *Electro-analytical Chemistry, Volume 22*, ed. A. J. Bard and I. Rubinstein, Marcel Dekker, New York, 2003, p. 101.
9. *Bioelectrochemistry – Fundamentals, Experimental Techniques and Applications*, ed. P. N. Bartlett, John Wiley & Sons, Chichester, 2008.
10. M. D. Newton and J. F. Smalley, *Phys. Chem. Chem. Phys.*, 2007, **9**, 555.

CHAPTER 5

More Complex Electrode Reactions

5.1 INTRODUCTION

While it has been recognized in earlier chapters that most electrode reactions are multistep sequences, so far the quantitative discussion (the Butler–Volmer equation, the Tafel equation and related expressions) has dealt only with electrode reactions that involve the simultaneous transfer of ne^-. Indeed, it can be questioned (see below) whether these equations are appropriate for reactions where $n > 1$. Certainly, they are not the full story for reactions such as hydrogen evolution:

$$2H^+ + 2e^- \longrightarrow H_2 \qquad (5.1)$$

the reduction of oxygen:

$$O_2 + 4H^+ + 4e^- \longrightarrow 2H_2O \qquad (5.2)$$

or the hydrodimerization of acrylonitrile:

$$CH_2 = CHCN + 2H_2O + 2e^- \longrightarrow (CH_2CH_2CN)_2 + 2OH^- \qquad (5.3)$$

that involve multiple electron transfers, bond cleavage/formation, protonations and perhaps adsorbed intermediates. This chapter will therefore outline approaches to handling more complex reactions. Following a brief discussion of multiple electron transfer reactions, the role of adsorbed species in electrode reactions will be developed. In particular, it will be shown that the involvement of adsorbed

A First Course in Electrode Processes, 2nd Edition
By Derek Pletcher
Published by the Royal Society of Chemistry, www.rsc.org

intermediates can lead to new reaction pathways with much higher rates – this is the basis for electrocatalysis.

One approach to studying complex reactions is the analysis of steady state current density, j, *versus* potential, E, data. Generally, plots of $\log j$ *vs* E have a linear range, *i.e.* the data follow the equation:

$$\log j = A + \frac{1}{B} E \qquad (5.4)$$

where A and B are constants. When analysing experimental data, however, it should be stressed that a Tafel slope is meaningful only if the $\log j$ *vs* E is linear over a significant current density and potential range, ideally a factor of 100 in current density. While Equation (5.4) has a similar form to the Tafel equation and confirms that the rates of most electrode reactions increase exponentially with applied potential, the interpretation of the constants is different. Firstly, the Tafel slope, B, must be interpreted in terms of the mechanism for the whole reaction and is not simply a reflection of the transfer coefficient, α. Experimental Tafel slopes can routinely vary between $(30\,\text{mV})^{-1}$ and $(>240\,\text{mV})^{-1}$. Secondly, the constant, A, has no fundamental interpretation. It can be interpreted in terms of the exchange current density, j_0, only if the equilibrium potential for the reaction is known and the data can be replotted as a function of overpotential. This is not always possible and extrapolation to 0 V *vs* a reference electrode leads to a current density that can be useful for comparing the rates of a reaction under different conditions, *e.g.* different electrode materials (provided the mechanism remains the same) but it cannot be used to give any insight into the mechanism.

5.2 MULTIPLE ELECTRON TRANSFER REACTIONS

Some electrode reactions involve the overall transfer of more than one electron but without the involvement of coupled chemical reactions and without clearly identifiable, intermediate oxidation states. The question then becomes – 'Are the electrons transferred simultaneously or in a sequence of $1e^-$ steps?'

Here, the particular example of the reaction:

$$Cu^{2+} + 2e^- \rightleftarrows Cu \qquad (5.5)$$

in acidic sulfate solution will be used to illustrate the general approach to answering this question. In contrast to chloride media where Cu(I) is a stable oxidation state and voltammograms for the reduction of Cu(II) show two, well separated, $1e^-$ reduction waves, Cu(I) is not recognized as a stable oxidation state in many media, *e.g.* sulfate and perchlorate solutions. In such media, voltammograms show a single $2e^-$ reduction wave. If the two electrons are transferred simultaneously, it is to be expected that the oxidation and reduction reactions will obey the Tafel equations:

$$\log j = \log j_o + \frac{\alpha_a nF}{2.3RT}\eta \tag{5.6}$$

and:

$$\log -j = \log j_o - \frac{\alpha_c nF}{2.3RT}\eta \tag{5.7}$$

respectively with, $n=2$ and $\alpha_a=\alpha_c=0.5$. Hence Tafel plots for both the reduction of Cu^{2+} and the oxidation of copper will give slopes of $(60\,mV)^{-1}$. Alternatively, if the reduction takes place in sequential $1e^-$ steps:

$$Cu^{2+} + e^- \rightleftharpoons Cu^+ \tag{5.8}$$

$$Cu^+ + e^- \rightleftharpoons Cu \tag{5.9}$$

the form of the Tafel plot will be different. One of the steps must be the slower and rate-determining step. If it is assumed that Reaction (5.9) is fast compared to (5.8), the rate-determining step is Reaction (5.8). The overall rate of conversion of Cu^{2+} into Cu metal is determined by the rate of a reaction in which $1e^-$ is transferred and the Tafel slope for the reduction will be $(120\,mV)^{-1}$, assuming that $\alpha=0.5$. The rate of oxidation of Cu back into Cu^{2+} must be determined by the kinetics of the same reaction. As described in Equations (1.36) and (1.38), the current density for the oxidation step (5.8) is given by:

$$j = Fk_a(c_{Cu^+})_{x=0} \tag{5.10}$$

where k_a is a potential dependent rate constant given by:

$$k_a = k_a^o \exp\frac{\alpha_a FE}{RT} \tag{5.11}$$

and c_{Cu^+} is the concentration of Cu^+ at the electrode surface. In chemical kinetics, when a rate-determining step follows a rapid one, the concentration of the intermediate, here Cu^+, is found either (a) by applying the steady state approximation or (b) by assuming that the rapid step is in equilibrium. Although less precise, the latter approach is simpler since the concentration can be readily estimated from the Nernst equation for the Cu^+/Cu couple:

$$E = E_e^o + \frac{RT}{F} \ln c_{Cu^+} \tag{5.12}$$

Substituting Equations (5.11) and (5.12) into (5.10) gives:

$$j = Fk_a^o \exp\frac{\alpha_a FE}{RT}\exp\frac{F(E - E_e^o)}{RT} \tag{5.13}$$

or:

$$\log j = \text{constant} + \frac{(1 + \alpha_a)F}{2.3RT}E \tag{5.14}$$

where the constant combines all the quantities independent of the experimental potential. Now, if $\alpha_a=0.5$, the Tafel slope will $(40\,\mathrm{mV})^{-1}$. Hence if Reaction (5.8) is the rate-determining step, the Tafel slopes for reduction and oxidation will be $(120\,\mathrm{mV})^{-1}$ and $(40\,\mathrm{mV})^{-1}$, respectively. An exactly analogous argument will show that if Reaction (5.9) is the slow step, the Tafel slopes for reduction and oxidation will be $(40\,\mathrm{mV})^{-1}$ and $(120\,\mathrm{mV})^{-1}$, respectively.

Hence, as summarized in Table 5.1, determining the Tafel slopes for both oxidation and reduction permits distinction between the mechanisms as well as the possible rate-determining steps. The experimentally determined Tafel slopes for reduction and oxidation are $(120\,\mathrm{mV})^{-1}$ and $(40\,\mathrm{mV})^{-1}$, respectively, showing that the reduction of Cu^{2+} to Cu occurs by a sequence of two discrete $1e^-$ steps with the first as the rate-determining step.

5.3   HYDROGEN EVOLUTION AND OXIDATION REACTIONS

The study of the hydrogen evolution and oxidation reactions has been central to the development of modern concepts of electrochemistry. In addition, they are reactions of unique importance in electrochemical

Table 5.1   Tafel slopes for the mechanisms possible for the reduction of Cu^{2+} to Cu.

Mechanism	Cathodic Tafel slope	Anodic Tafel slope
Simultaneous $2e^-$ transfer	$(60\,\mathrm{mV})^{-1}$	$(60\,\mathrm{mV})^{-1}$
Sequential $2 \times 1e^-$ steps – first $1e^-$ transfer as rate-determining step	$(120\,\mathrm{mV})^{-1}$	$(40\,\mathrm{mV})^{-1}$
Sequential $2 \times 1e^-$ steps – second $1e^-$ transfer as rate-determining step	$(40\,\mathrm{mV})^{-1}$	$(120\,\mathrm{mV})^{-1}$

technology. Hydrogen oxidation is a reaction in H_2/O_2 fuel cells while hydrogen evolution is the cathode reaction in water electrolysis, membrane and diaphragm cells in the chlor-alkali industry as well as a convenient counter electrode process in anodic electrosyntheses and environmental technology involving anodic oxidation. In such situations, the goal is a surface where the kinetics of the H^+/H_2 couple are rapid, *i.e.* the surfaces are electrocatalysts. In contrast, hydrogen evolution is an inevitable but unwanted competing reaction in electrowinning, electroplating and cathodic electrosyntheses and environmental technology involving cathodic reduction. Then, the interest is in surfaces where the kinetics of the H^+/H_2 couple are slow. Hydrogen evolution is also commonly the cathodic process in the corrosion of metals and diminishing the kinetics of the hydrogen evolution reaction is one approach to protecting the metal. Finally, the H^+/H_2 couple has a special importance in thermodynamics; the equilibrium potential of the couple in an aqueous solution of protons (concentration, 1 M) is, by convention, taken to be zero. Under these circumstances, it is not surprising that the literature contains a large body of data about the reactions and many surfaces have been investigated.

Here, the discussion of hydrogen evolution will be developed assuming an aqueous acid solution where the overall reaction is:

$$2H^+ + 2e^- \rightleftharpoons H_2 \qquad (5.15)$$

Modification of the equations to neutral and alkaline media where the reaction becomes:

$$2H_2O + 2e^- \rightleftharpoons H_2 + 2OH^- \qquad (5.16)$$

is straightforward. The key intermediate in the hydrogen evolution reaction is the adsorbed hydrogen atom formed in the reaction:

$$M + H^+ + e^- \longrightarrow M - H \qquad (5.17)$$

where M is a site (*e.g.* a metal atom) on the surface of a metal cathode. The formation of an adsorbed hydrogen atom changes the Gibbs free energy for the reduction of a proton by an amount equal to the Gibbs free energy of adsorption of the hydrogen atom. This stabilization of the product of electron transfer is seen as a positive shift in the potential for the reduction of the protons and this shift will be equal to $-\Delta G_{ADS}/F$ (where ΔG_{ADS} is a negative quantity). The extent of this reaction will be limited by the number of surface sites available; at most a monolayer of adsorbed H atoms can be formed. Clearly, the number of surface sites and the Gibbs free energy of adsorption will be determined by the cathode material and its form. The cathodic adsorption of hydrogen has been particularly well characterized at the platinum metals, using, for example, cyclic voltammetry, and the behaviour at Pt surfaces is discussed in Chapters 2, 6 and 7.

In terms of the hydrogen evolution reaction (and also hydrogen oxidation), it is the additional pathways for the interconversion of protons and hydrogen gas opened up by the existence of adsorbed H atoms that is important in electrocatalysis. If such pathways have a lower energy of activation, then the kinetics will be enhanced; a higher current density will be observed at each overpotential and a lower overpotential is necessary to achieve any demanded current density. Two pathways for hydrogen evolution involving adsorbed H atoms have been generally considered:

(I)

$$M + H^+ + e^- \longrightarrow M - H \tag{A}$$
$$2M - H \longrightarrow H_2 + 2M \tag{B}$$

(II)

$$M + H^+ + e^- \longrightarrow M - H \tag{A}$$
$$M - H + H^+ + e^- \longrightarrow H_2 + M \tag{C}$$

and in each pathway either the first or second step can be the rate-determining step. Mechanism I envisages that the formation of the adsorbed hydrogen atom is followed by the combination of two such entities and the release of H_2 gas (thereby making available two surface sites for further chemical change). In mechanism II, the second step is discharge of a proton at a surface site already covered by an adsorbed H atom.

Notably, mechanisms I or II require both the formation and then cleavage of M–H bonds. In consequence, it can be predicted that the best electrocatalysts will always have surfaces that give M–H bonds of intermediate strength. If the Gibbs free energy of adsorption is too small the coverage by adsorbed H atoms will be too low, while if it is too large the first step will occur well but lead only to a monolayer of adsorbed H atoms as the second steps will be retarded. The desired situation is where there is a significant coverage by adsorbed H atoms but the second step

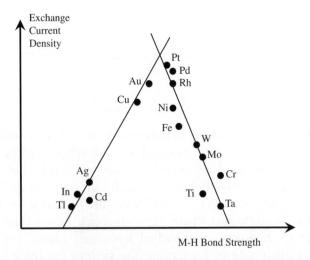

Figure 5.1 Variation of exchange current density for the H_2 evolution reaction at various metals as a function of the M–H bond strength.

is also facile. Such behaviour is indeed observed and Figure 5.1 shows a plot of the exchange current density for the hydrogen evolution reaction *vs* Gibbs free energy of adsorption of H atoms for several metals. The maximum at intermediate Gibbs free energies of adsorption is obvious. Such plots are common in heterogeneous gas-phase catalysis and have been termed 'volcano plots'.

The Tafel slopes and also the reaction orders can be derived for each of the possible mechanisms. Here, we consider H_2 evolution and, for simplicity, it will be assumed that (a) at equilibrium, the surface coverage by adsorbed hydrogen atoms may be estimated from the Langmuir isotherm (similar derivations are possible with other isotherms but the algebra is more complex) and (b) the rate-determining steps are irreversible; for an electron transfer reaction, this means that a significant overpotential is being applied and, therefore, hydrogen oxidation can be ignored. The expressions for H_2 oxidation can be derived in a similar way but experimentally this reaction is much more difficult to study because of the low solubility of hydrogen gas in aqueous solutions.

5.3.1 Mechanism I or II – Reaction A as the Rate-determining Step

The formation of adsorbed hydrogen atoms is common to both mechanisms (I) and (II). Hence, under conditions where step A determines the kinetics of H_2 evolution, it is not possible to determine which mechanism occurs. It is possible only to recognize that step A is the rate-determining step.

The rate of step A (and hence the rate of H_2 evolution when it is the slow step) may be written:

$$\text{Rate of reaction A} = k_A c_{H^+}(1 - \theta) \tag{5.18}$$

where k_A is the potential dependent rate constant for reaction A and θ is the fraction of the surface covered by adsorbed hydrogen atoms. Equation (5.18) assumes that the rate of reaction A is first order in both reactants, the proton in solution and the fraction of the surface free of adsorbed hydrogen atoms $(1 - \theta)$ and therefore available for reaction A to take place. In situations where step A is much slower than steps B or C, the rate of removal of adsorbed hydrogen atoms will be rapid compared to their rate of formation; the surface coverage will be low and $(1 - \theta)$ tends to one. A very simple expression may be written for the current density for H_2 evolution:

$$\begin{aligned}
-j &= F k_A c_{H^+} \\
&= F c_{H^+} k_A^{\circ} \exp -\frac{\alpha F}{RT} E
\end{aligned} \tag{5.19}$$

since the rate constant will have a normal potential dependence. Taking logarithms:

$$\log -j = \text{constant} + \log c_{H^+} - \frac{\alpha F}{2.3 RT} E \tag{5.20}$$

leads to an expression with a very similar form to the Tafel equation for a simple electrode reaction (note the current density, j, is a negative quantity for a cathodic reaction and hence Equation (5.20) requires the logarithm of a positive quantity). It can be seen that the current density is first order in proton (provided it is measured at a constant potential *vs* a reference electrode) and if $\alpha = 0.5$ the Tafel slope is $(120 \, \text{mV})^{-1}$.

5.3.2 Mechanism I – Reaction B as the Rate-determining Step

The rate of reaction B is given by:

$$\text{Rate of reaction B} = 2 k_B \theta^2 \tag{5.21}$$

where k_B is a chemical rate constant (therefore independent of potential). The factor of two recognizes that two adsorbed hydrogen atoms are consumed each time the reaction takes place once. The fractional surface coverage may be found by noting that in the steady state:

$$\frac{d\theta}{dt} = k_A c_{H^+}(1 - \theta) - k_{-A}\theta - k_B\theta^2 = 0 \tag{5.22}$$

This expression is somewhat intractable but under a range of experimental conditions where reaction B is the rate-determining step the final term can be neglected, *i.e.* the rate of reaction B is slow compared to both forward and back reactions for step A. Then algebra leads to:

$$\theta = \frac{k_A c_{H^+}}{k_{-A} + k_A c_{H^+}}$$
$$= \frac{\frac{k_A^{\circ}}{k_{-A}^{\circ}} c_{H^+} \exp - \frac{F}{RT} E}{1 + \frac{k_A^{\circ}}{k_{-A}^{\circ}} c_{H^+} \exp - \frac{F}{RT} E} \tag{5.23}$$

after introducing the potential dependences of k_A and k_{-A} and simplifying the expression. This equation has two limiting forms:

1. At high negative potentials: $(k_A^{\circ}/k_{-A}^{\circ})c_{H^+} \exp -(F/RT)E > 1$, when $\theta \rightarrow 1$ and the current density becomes independent of potential.
2. At less negative potentials: $(k_A^{\circ}/k_{-A}^{\circ})c_{H^+} \exp -(F/RT)E < 1$, when:

$$\theta = \frac{k_A^{\circ}}{k_{-A}^{\circ}} c_{H^+} \exp - \frac{F}{RT} E \tag{5.24}$$

Substituting this into Equation (5.21) and taking logarithms leads to:

$$\log -j = \text{constant} + 2\log c_{H^+} - \frac{2F}{2.3RT} E \tag{5.25}$$

In this range of potentials, hydrogen evolution with step B as the rate-determining step is second order in proton and will have a Tafel slope of $(30\,\text{mV})^{-1}$.

5.3.3 Mechanism II – Reaction C as the Rate-determining Step

The rate of reaction C can be written:

$$\text{Rate of reaction C} = k_C c_{H^+}\theta \tag{5.26}$$

since it depends on the concentration of protons in solution and the fraction of the surface area covered by adsorbed hydrogen atoms (and therefore available for reaction C to occur). Again, k_C is a potential-dependent rate constant because reaction C involves electron transfer. The expression for the surface coverage, θ, again results from writing an equation for $d\theta/dt$ and considering the steady state. For the situation where reaction C is slow compared to reaction A and at high over-potentials, $\theta \rightarrow 1$ and at low overpotentials the fractional coverage is again given by (5.24). The two limiting situations are:

1. At high negative potentials, $\theta \rightarrow 1$ and substituting for the potential-dependent rate constant:

$$-j = Fk_C^o c_{H^+} \exp -\frac{\alpha F}{RT} E \qquad (5.27)$$

and:

$$\log -j = \text{constant} + \log c_{H^+} - \frac{\alpha F}{2.3RT} E \qquad (5.28)$$

In this range of potentials, hydrogen evolution with step C as the rate-determining step is first order in proton and will have a Tafel slope of $(120 \, \text{mV})^{-1}$.

2. At lower negative potentials:

$$-j = F\left(k_C^o \exp -\frac{\alpha F}{RT} E\right) c_{H^+} \left(\frac{k_A^o}{k_{-A}^o} c_{H^+} \exp +\frac{F}{RT} E\right) \qquad (5.29)$$

and collecting together like terms and taking logarithms:

$$\log -j = \text{constant} + 2\log c_{H^+} - \frac{(1+\alpha)F}{2.3RT} E \qquad (5.30)$$

In this range of potentials, hydrogen evolution with step C as the rate-determining step is second order in proton and will have a Tafel slope of $(40 \, \text{mV})^{-1}$, assuming $\alpha = 0.5$.

The change in slope from $(40 \, \text{mV})^{-1}$ at low overpotentials to $(120 \, \text{mV})^{-1}$ at high overpotentials can be seen experimentally at some Pt surfaces.

Evidently, determination of the Tafel slopes and the reaction orders with respect to protons gives an insight into the mechanism; Table 5.2 summarizes conclusions for mechanisms I and II. The values for the reaction order and Tafel slope are, however, not unique to one

Table 5.2 Tafel slopes and reaction orders predicted for four mechanisms for the hydrogen evolution reaction.

Mechanism	Rate-determining step	Overpotential range	Tafel slope $\left(\frac{d\log -j}{dE}\right)$	Reaction order $\left(\frac{d\log -j}{d\log c_{H^+}}\right)$
I or II	A	All	$(120\,mV)^{-1}$	1
I	B	Low	$(30\,mV)^{-1}$	2
II	C	Low	$(40\,mV)^{-1}$	2

Table 5.3 Exchange current densities and Tafel slopes for the H_2 evolution reaction in 1 M H_2SO_4 at 298 K. Data taken from A. J. Appleby, M. Chemla, H. Kita and G. Bronoël in *Encylcopedia of the Electrochemistry of the Elements*, Volume IXA, ed. A. J. Bard, Marcel Dekker, New York 1982, p. 383.

Metal	$Log -j_o/mA\ cm^{-2}$	Tafel slope$/mV^{-1}$
Ag	−3.4	120
Au	−2.7	120
Cd	−9.0	120
Co	−1.9	120
Cr	−3.4	120
Cu	−4.4	120
Fe	−2.8	120
Hg	−8.9	120
Ir	−0.3	120
Nb	−4.3	120
Ni	−2.2	120
Pb	−9.6	120
Pd	+0.6	40
Pt	−0.3	30
Rh	+0.5	60
Ru	−0.3	120
Sn	−6.2	120
Ti	−3.9	120
W	−3.4	60
Zn	−7.5	120

mechanism (and other values are possible if other mechanisms or adsorption isotherms are considered) and hence other experimental data are essential to identify the mechanism with certainty. As noted above, it should also be recognized that a Tafel slope is meaningful only if the log j vs E plot is linear over a significant current density and potential range, ideally a factor of 100 in current density.

Steady state current density vs potential data are available for numerous cathode materials. Table 5.3 reports estimated values for the Tafel slopes and exchange current densities for H_2 evolution. It should be emphasized that the exact values will be a function of the state of the metal surfaces and hence the pre-treatments of the surface before the data are recorded. Also, the Tafel slopes reported have been rounded to likely theoretical values. The data are intended to stress the large variation in the exchange current densities – in fact ten orders of magnitude! Furthermore, it can be seen that for many metals the first electron transfer, step A is the rate-determining step. The exceptions are all good electrocatalysts, confirming the importance of adsorbed H atoms in the electrocatalysis of this reaction. In practice, the current density at any overpotential is a function of the Tafel slope as well as the exchange current density; when values of current are obtained by extrapolation over a significant potential difference, small errors in the Tafel slope can lead to poor estimates of the current density. Indeed, when reactions are studied at potentials removed from the equilibrium potential, uncertainties in the exchange current density are inevitable.

It should also be reiterated that the large variation in catalytic activity is highly beneficial to electrochemical technology. When H_2 evolution or oxidation is the goal, the Pt metals provide good catalysis. Mercury, cadmium and lead are the classical examples of high hydrogen over-potential materials suitable for cathodes when H_2 evolution is an unwanted competing reaction.

5.4   OXYGEN EVOLUTION AND REDUCTION

Oxygen reduction is a significantly more complex reaction. Firstly, the product may be hydrogen peroxide formed in a $2e^-$ reduction:

$$O_2 + 2H^+ + 2e^- \rightleftharpoons H_2O_2 \qquad (5.31)$$

with a formal potential of $+0.69$ V *vs* SHE or water formed in a $4e^-$ reduction:

$$O_2 + 4H^+ + 4e^- \rightleftharpoons H_2O \qquad (5.32)$$

with a formal potential of $+1.23$ V *vs* SHE. Moreover, in particular the $4e^-$ reduction is clearly a multistep sequence that involves $4e^-$ and both bond formation and bond cleavage steps.

It is generally accepted that the $4e^-$ reduction can occur via two types of reaction pathways, pathways A and B, set out in Figure 5.2. Pathway A involves formation of water through a route where the O–O bond is cleaved early in the reaction sequence. Figure 5.3 illustrates this type of

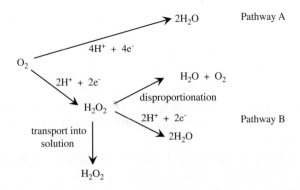

Figure 5.2 Reaction pathways for the reduction of oxygen in acid solution.

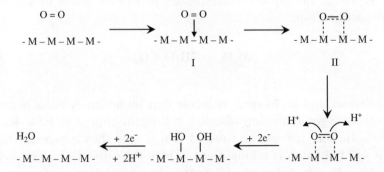

Figure 5.3 One mechanism for the 4e⁻ reduction of oxygen in a sequence involving the formation of a π-complex between O_2 and the surface (I), cleavage of the O–O bond with concerted formation of M–O bonds (II), and protonation and reduction to form water.

reaction pathway. Transport of the oxygen to the surface is followed by the formation of a π-complex between the oxygen molecule and the surface and the O–O bond is cleaved through a concerted mechanism, *i.e.* surface–oxygen bonds are formed simultaneously with weakening of the O–O bond. Protonation and reduction leads to M–OH bonds that are then reduced further. In such a pathway, the electrode material will have a key role in determining the kinetics of the sequence. Firstly, the formation of the π-complex is dependent on the availability of suitable energy levels to receive the donated electrons. Then the length of M–M bonds (or more generally the spacing between active sites) as well as the electronic energy levels in the electrode material will determine the probability and rate of the concerted steps, including the protonation

steps. Finally, it must be possible to cleave the M–OH bonds by reduction. The active sites have several different roles to play to be efficient
catalysts for the $4e^-$ reduction; one way for them to achieve such diverse
chemistry is to be able to change oxidation state. Unsurprisingly,
therefore, the effective catalysts are transition metals or an oxide or
complex containing a transition metal ion. It is possible to model the
surface by considering the lattice as moderating the electronic properties
of the metal centre.

Pathway B for the reduction of oxygen considers hydrogen peroxide
as a discrete intermediate (Figure 5.2). Indeed, under some conditions
and with some electrode materials, hydrogen peroxide is the only
product formed and then it is possible to produce solutions of hydrogen
peroxide in high yield. At other surfaces, the hydrogen peroxide will
undergo either (a) rapid further reduction to give water or (b) rapid
disproportionation:

$$2H_2O_2 \rightarrow 2H_2O + O_2 \tag{5.33}$$

to give water and an oxygen molecule that immediately reduces again.
Then the oxygen reduction reaction will again appear to be a $4e^-$ reduction. Indeed, it must be stressed that if it is shown experimentally
that a $4e^-$ reduction is taking place, it is not possible to know whether
pathway A or pathway B (with rapid further reduction or
disproportionation of hydrogen peroxide) is followed. Certainly, in both
cases, the thermodynamics and hence the equilibrium potential will
be determined by the O_2/H_2O couple. At many electrode materials, a
relatively small amount of hydrogen peroxide is observed (*e.g.* using
a RRDE experiment, see Chapter 7); it is then, again, difficult to
demonstrate conclusively whether one is observing (a) competition
between pathways A and B or (b) pathway B with competition between
transport of hydrogen peroxide away from the surface and conversion
into water at the surface.

Analysis of possible reaction mechanisms by a procedure analogous
to that illustrated for hydrogen evolution in the previous section is
possible. Consideration of the reaction order with respect to oxygen,
the influence of pH and the Tafel slope will allow insight into the
mechanism at each electrode material. However, the procedure is more
difficult and time consuming. There are substantially more steps to be
considered, the algebra becomes more complex and it is necessary
to consider more limiting cases. Such an analysis will therefore not be
attempted here.

5.5 OTHER REACTIONS

Many other important electrode reactions are dependent on adsorbed intermediates to achieve a significant current density.

While most fuel cell development is focused on H_2/O_2 fuel cells (Chapter 8), it would be highly beneficial to have effective electrocatalysts for the oxidation of organic molecules and hence the ability to operate fuel cells with organic, even hydrocarbon, fuels without a converter. There is development of methanol fuel cells where the objective is to carry out the complete oxidation of the methanol to carbon dioxide:

$$CH_3OH + H_2O - 6e^- \longrightarrow CO_2 + 6H^+ \qquad (5.34)$$

(to release the full energy stored in the methanol molecule) at a high current density close to the equilibrium potential ($\sim +30\,mV$ *vs* SHE). Reaction (5.35) does not provide a viable first step as it occurs only at very high overpotentials:

$$CH_3OH - e^- \longrightarrow CH_3OH^+ \qquad (5.35)$$

Hence, it is essential to design surfaces (*e.g.* precious metal surfaces) that sustain alternative pathways, for example, routes where the first step is:

$$CH_3OH + 2M \longrightarrow M-H + M-CH_2OH \qquad (5.36)$$

This initial step requires the cleavage of a strong bond but once achieved it could be followed by facile oxidation of adsorbed H atom and cleavage of further C–H bonds. Another problem with the oxidation of organic molecules is poisoning by carbon monoxide. Repetitive disassociation of the C–H bonds of methanol would lead to adsorbed carbon monoxide and carbon monoxide is very strongly bound to precious metal surfaces; it is difficult either to oxidize or desorb. Possible routes to the oxidation of both organic fuels and carbon monoxide involve the formation of an oxidizing agent on the surface of the electrode. For example, a mechanism for the oxidation of carbon monoxide on a platinum surface would be:

$$Pt + H_2O - e^- \longrightarrow Pt-OH + H^+ \qquad (5.37)$$

$$Pt-OH + Pt-CO \xrightarrow{-e} CO_2 + 2Pt + H^+ \qquad (5.38)$$

where the 'OH' species and CO are on neighbouring sites on the surface. Such mechanisms are one reason for employing alloy catalysts since it can be envisaged that the two metals have different roles in the oxidation. For example, as ruthenium is more easily oxidized than platinum the use of a PtRu alloy introduces the possibility of more facile formation of a 'OH' species on the surface.

The electrosynthesis of organic molecules can also use pathways where absorbed intermediates are critical, thereby introducing greater selectivity. For example, unsaturated molecules (*e.g.* alkynes, alkenes, ketones, aldehydes, nitriles) can frequently be hydrogenated by co-adsorption of the organic compound with hydrogen atoms, *i.e.* using Pt, Pd or Ni cathodes at potentials where adsorbed hydrogen atoms are formed by electron transfer at the surface:

$$2M\text{-}H \ + \ \left(\!\!\begin{array}{c}\diagdown\\C\!=\!X\\\diagup\end{array}\!\!\right)_{ads} \ \longrightarrow \ 2M \ + \ \begin{array}{c}\diagdown\\C\text{-}XH\\\diagup\end{array} \tag{5.39}$$

This is the electrochemical version of heterogeneous catalysis. Under other conditions, the same transformation might be triggered by electron transfer:

$$\begin{array}{c}\diagdown\\C\!=\!X\\\diagup\end{array} + \ e^- \ \longrightarrow \ \begin{array}{c}\diagdown\\C\!=\!X^{-\cdot}\\\diagup\end{array} \tag{5.40}$$

but usually this would be at a potential 1–2 V more negative and the anion radical is a highly reactive intermediate with alternative decomposition routes, leading to a less selective conversion.

5.6   ELECTROCATALYSIS

The aim of electrocatalysis is to design surfaces so as to obtain a high current density (*i.e.* high rate of conversion) close to the equilibrium potential (*i.e.* at low overpotential). Some electrode reactions are reversible (*i.e.* they have large standard rate constants) at a wide range of electrode materials and, hence, have no need for a catalyst. In contrast, many other electrode reactions require a substantial overpotential to drive them at a practical current density. In these cases, it may be that a new, low energy of activation route from reactant to product can be created by designing an appropriate surface that stabilizes an adsorbed species to an optimum extent. Such electrocatalytic reactions have several characteristics:

- The kinetics and perhaps the mechanism (*i.e.* exchange current density and Tafel slope) depend strongly on the choice of electrode material – certainly on its chemical composition but perhaps also on its form. Materials that allow practical current densities at low over-potentials are termed electrocatalysts. As well as metals, electrocatalysts can commonly be an alloy, a metal oxide or a transition metal complex (either adsorbed or bonded to the substrate).

- Since the current density reflects the total number of (active) surface sites exposed to the electrolyte and available for the electrode reaction, the current density based on the apparent geometric area can be much enhanced by the deliberate preparation of a high area interface by roughening the surface or dispersing the electrocatalyst as small centres on an inert substrate. This is in contrast to mass transport controlled reactions when the current density depends only on the apparent geometric area.

- However good an electrocatalyst may be, the rate of reaction is still subject to mass transport limitations – the mass transport controlled current density is always the maximum rate of chemical change. A consequence has been the development of gas diffusion electrodes (GDE) for gaseous reactants with a low solubility in the electrolyte (Chapter 8).

- Selectivity may also be an issue. For example, in chlorine production the anodes must catalyse Cl_2 evolution but also inhibit O_2 evolution, the thermodynamically preferred reaction. The anode materials are judged both by their overpotential and current efficiency for Cl_2 evolution.

- Electrocatalysts are often poisoned by the adsorption of unwanted species (*e.g.* impurities in the electrolyte) on the surface.

In modelling or understanding electrocatalytic reactions, a critical question is whether the electrocatalytic reaction takes place over all the surface or whether it occurs only at special sites on the surface (*e.g.* step sites, see Figure 2.2). Certainly, it is well established that the rate of electrocatalytic reactions can depend on the crystal face exposed to the solution (see Figure 2.3). Although the original reason for using dispersed catalyst on a substrate was to reduce the cost of the catalyst, there is now evidence that, on a 1–10 nm scale, the catalytic activity depends on the size of the centres. In addition, the selection of the substrate can also influence catalytic activity and, although a high area carbon remains the most common material, stable non-stoichiometric oxides such as WO_{3-x} or TiO_{2-x} are growing in importance.

Electrocatalysts have an increasing role in electrochemical technology; it is, however, important to recognize that practical electrocatalysts must meet more criteria than giving a high current density at low overpotential. Table 5.4 sets out the main characteristics sought.

Table 5.4 Properties sought for practical electrocatalysts/substrates.

1. A high current density at low overpotential
2. No competing reactions
3. Availability as a high area coating on a substrate suitable
   for fabricating electrodes of diverse shape
4. Stable (to corrosion *etc*) on load, off-line and during switching
5. High mechanical stability
6. Performance maintained over long periods (years?)
7. Acceptable cost

Table 5.5 lists some of the electrocatalysts established in electrochemical technology. They fall into two types. Those developed for fuel cells are usually a precious metal dispersed over a high surface area form of carbon and fabricated into gas diffusion electrodes and these structures will be discussed further in Chapter 8. Here it should just be noted that currents $> 1\,A\,cm^{-2}$ for both hydrogen oxidation and oxygen reduction can be achieved with metal loadings of $\sim 0.1\,mg\,cm^{-2}$ of geometric area and without a mass transport limitation. Those intended for electrolytic cells are usually coatings on titanium, steel or nickel and these may be plates or shaped electrodes. Techniques for preparing the coatings include spraying of a solution containing a precursor followed by a thermal treatment, vacuum sputtering and, occasionally, electroplating.

Perhaps the most successful electrocatalysts are the so-called dimensionally stable anodes (or DSA) developed for the generation of chlorine. When deposited as a layer a few microns thick on titanium, the ruthenium dioxide based formulations have an operational lifetime of > 5 years. The overpotential is $< 100\,mV$ for a current density of $0.5\,A\,cm^{-2}$ and the current efficiency for Cl_2 is $> 99.9\%$ despite O_2 evolution being the thermodynamically preferred reaction in the electrolysis medium, typically 25 wt% NaCl at pH 2. There are also satisfactory electrocatalyst materials for O_2 and H_2 evolution although the recommended material does depend on the electrolysis medium. Few other electrode reactions merit the cost of developing electrocatalysts specifically for the one reaction and hence the choice of electrode material is usually based on experience.

Even now, electrocatalysts are still being developed and improved using 'guiding principles' and largely empirical approaches. The general concepts are clear:

- the dominant role of adsorbed intermediates;
- selecting the catalyst material and using the environment of the surface atoms to tailor the electronic levels within the catalyst to give the optimum strength of adsorption;

Table 5.5 Some widely used electrocatalysts for reactions in electrochemical technology.

Electrode reaction	Medium	Catalyst	Substrate	Comment	Application
H_2 evolution	Strong base	Raney® type Ni, Ni alloys, Pt metals	Ni or steel.		Cathode in Cl_2/NaOH industry. General cathode reaction
	Acid	Pt	Ti		General cathode reaction
H_2 oxidation	Acid Nafion™	Pt, Pt alloys	High area C powder	Highly dispersed catalyst	Fuel cell anode
	Base	Ni, Ni alloys	High area C powder	Highly dispersed catalyst	Fuel cell anode
O_2 evolution	Acid	IrO_2, PbO_2, Pt	Ti, Ti, C, Ti	Known as DSA O_2, Stable but high η	Water electrolysis, General anode reaction, Water electrolysis
	Strong base	Ni, Raney® type Ni, $NiCo_2O_4$	Ni or stainless steel		
O_2 reduction	Acid Nafion™	Pt, Pt alloys	High area C powder	Highly dispersed catalyst	Fuel cell cathode
	Base	Pt, Ni, MnO_2	High area C powder	Highly dispersed catalyst	Highly dispersed catalyst
Cl_2 evolution	Concentrated NaCl, pH 2	RuO_2	Ti	Known as DSA Cl_2	Anode in Cl_2/NaOH industry

- recognizing the importance of particular surface sites;
- geometric factors such as the spacing of active sites and the importance of creating sites where nearest neighbours are different atoms;
- creating surfaces with the highest number of active sites/unit area and thereby introducing 'cost efficiency' into the use of the electrocatalyst.
- However, the precise application of these principles to particular reactions remains beyond our capabilities.

Further Reading

1. J. O'M. Bockris and A. K. N. Reddy, *Modern Electrochemistry, Volume 2A – Fundamentals of Electrodics*, Kluwer Academic, New York, 2001.
2. J. O'M. Bockris and S. U. M. Khan, *Surface Electrochemistry – A Molecular Level Approach*, Plenum, New York, 1993.
3. *Handbook of Fuel Cells, Volume 2 'Electrocatalysis'*, ed. W. Vielstich, A. Lamm and H. A. Gasteiger, John Wiley & Sons, New York, 2003.
4. *Electrocatalysis*, ed. J. Lipkowski and P. N. Ross, Wiley-VCH Verlag GmbH, Weinheim, 1998.
5. S. Trasatti, in *Advances in Electrochemical Science and Engineering, Volume 2*, ed. H. Gerischer and C. W. Tobias, VCH, Weinheim, 1992.
6. K. Kinoshita, *Electrochemical Oxygen Technology*, John Wiley & Sons, New York, 1992.
7. D. Pletcher and F. C. Walsh, *Industrial Electrochemistry*, Chapman and Hall, London, 1990.

CHAPTER 6

Experimental Electrochemistry

6.1 INTRODUCTION

Obtaining high quality and reproducible data from electrochemical experiments requires some understanding of the problems faced and the approaches available to limit the extent to which they distort the data. Chapter 1 introduced the importance of designing the experiment so that the response can be understood and analysed quantitatively; for example, the importance of a defined mass transport regime was high-lighted. Critically important is the design of the electrochemical cell, and this will be a major component of this chapter. First, however, it is necessary to discuss some of the general problems met in the laboratory.

The objectives of the experiments should be clearly recognized. The most common would be (a) to define the overall chemical change brought about in the electrode reaction, (b) to identify the individual reaction steps and particularly the rate-determining step, (c) to determine the reaction mechanism and overall kinetics, (d) to optimize the rate and selectivity of the reaction of interest, (e) to carry out a clean chemical conversion and (f) to determine the concentration of a species in solution. All will require the recording of clean, undistorted responses and the interpretation of the data. We would stress the importance of observing and understanding the crude experimental data before any quantitative analysis is attempted. This warning is particularly important with the widespread application of PC-based control and data acquisition/analysis packages!

A First Course in Electrode Processes, 2nd Edition
By Derek Pletcher
© Derek Pletcher 2009
Published by the Royal Society of Chemistry, www.rsc.org

6.2 THE PROBLEMS

6.2.1 *IR* Drop

Forcing a current through a solution will always lead to a voltage drop equal to *IR*. Thus, when the current is passing between two electrodes, the potential difference will not be the applied voltage. The potential difference is given by:

$$\Delta E = \Delta V - IR_{\text{soln}} \tag{6.1}$$

Clearly, the magnitude of this unwanted *IR* drop will depend on both the current under the experimental conditions and the resistance of the solution between the electrodes, R_{soln}. The measured current density will depend on several factors, including the concentration of the electroactive species, the potential and the mass transport conditions (a function of potential scan rate, rotation rate, *etc.*) and in common laboratory experiments the cell current may range between 1 nA and 1 A. The magnitude may depend strongly on the exact experiment; for example, in cyclic voltammetry, the IR_{soln} drop will be a greater problem at high scan rates ($j_p \propto v^{1/2}$, see Chapter 7) and if a high concentration of electroactive species is employed. What is certain is that in most experiments (*e.g.* a voltammogram) the cell current will vary (and may go through a peak) and, in consequence, the IR_{soln} drop will vary, giving the possibility of a complex distortion of the response. The resistance will depend on the conductivity of the electrolyte solution (this is why aqueous acids or bases are favoured media for electrochemistry) and also the separation of the two electrodes (note that since $R_{\text{soln}} \propto 1/A$ and $j = AI$, the discussion can be in terms of current or current density). Depending on the cell geometry and dimensions, R_{soln} can easily vary between 0.1 and 1000 ohm. Hence, the unwanted *IR* drop can vary from totally insignificant to more than 1 V. The precise level when the distortion becomes 'significant' will depend on the accuracy in potential measurements demanded by the experiment.

It is the magnitude of the current density and the properties of the solution that determine our experimental approach to obtaining responses free from *IR* distortion. In experiments where the current is low and the medium highly conducting, it is possible to work with a two-electrode cell containing a working electrode and a reference electrode. It may even be possible to use a commercial reference electrode despite their high resistance. If the current is somewhat higher or the medium more resistive, it may still be possible to use a two-electrode cell if either a microelectrode is used as the working electrode or a low resistance

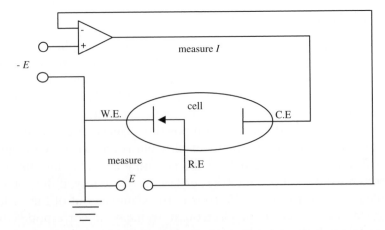

Figure 6.1   Concept of a circuit with a simple potentiostat and a three-electrode cell.

reference electrode is placed close to the working electrode. With much higher currents, it becomes necessary to employ a three-electrode cell and a potentiostat (Figure 6.1). In a three-electrode cell, the potential of the working electrode is controlled *versus* a reference electrode and the current flows between the working and a counter electrode. The potentiostat is a feedback circuit based on an operational amplifier circuit that seeks to ensure that the current in the working/reference electrode circuit approaches zero so that, in concept, the *IR* drop between the electrodes should be effectively zero. In reality, if the reference electrode is placed directly into the cell solution it probes the potential in the solution phase at the boundary with the internal solution of the reference electrode. If the reference electrode tip is close to the working electrode, the arrangement will lead to a significant reduction in the *IR* distortion. It is, however, preferable to mount the reference electrode within a Luggin capillary (see Section 6.5) and to place the tip of the Luggin capillary ~ 1 mm from the surface of the working elec-trode. Even then, there is an uncompensated resistance, R_u, for the slug of solution between the tip of the Luggin capillary and the electrode surface. Such a cell arrangement will generally allow the recording of distortion-free responses when the cell current is <20 mA cm^{-2}. Of course, the potentiostat shown in Figure 6.1 is very simple compared to commercial instruments and is intended only to indicate the concept. With higher current densities or highly resistive solutions it may be necessary to employ instrumental corrections or to use current interruption techniques. It is, however, advisable to avoid such procedures if at all possible; their correct application is beyond the scope of this book.

6.2.2 Double Layer Charging Currents

In Chapter 3 it was shown that, when a potential is applied to an electrode in an electrolyte solution, the electrode surface takes up a characteristic charge and attracts a layer of ions of opposite charge to form a double layer. The structure of the double layer, including the surface charge, is a function of the applied potential as well as the electrode material and solution composition. Hence, a consequence of changing the potential during any experiment is the necessity to change the surface charge (Figure 6.2). If the potential is made more negative, electrons must be pumped into the surface or, conversely, if the potential is made more positive then electrons must be moved out of the surface. Whichever the direction of electron movement, it corresponds to a current through the external surface. However, these electrons do not cross the electrode/solution interface and therefore do not bring about chemical change.

In consequence, the experimentally measured current density will have two components: (a) the Faradaic current density, $j_{Faradaic}$, corresponding to the electrons crossing the interface and leading to chemical change and (b) the non-Faradaic (or double layer charging) current density, $j_{charging}$, resulting from the electrons required to change the surface charge, *i.e.*:

$$j = j_{Faradaic} + j_{charging} \qquad (6.2)$$

In most experiments, it is desirable that we measure only the Faradaic component of the current. Notably, after a change in potential, the change of surface charge only has to occur once and it is also a fairly rapid process. Hence, in the steady state, there is no charging current component to the measured current and the charging currents will be most prevalent in short timescale experiments. More specifically, the time for the charging current to decay depends on both the double layer capacitance for the electrode/solution interface and the magnitude of the

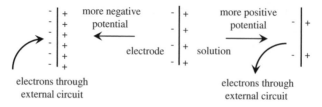

Figure 6.2 The origin of the charging current. The consequences of taking an electrode negative to the potential of zero charge and making the potential more negative or more positive.

Table 6.1 Comparison of double layer charging current densities ($j_{charging}$) and peak current densities (j_p), calculated using Equations (6.3) and (7.53), respectively, for a typical set of cyclic voltammograms; $C_{dl} = 20\ \mu F\ cm^{-2}$, $n = 1$, $c = 1\ mM$, $D = 6 \times 10^{-6}\ cm^2\ s^{-1}$.

Potential scan rate, $v/V s^{-1}$	$j_{charging}/mA\ cm^{-2}$	$j_p/mA\ cm^{-2}$
0.001	2×10^{-5}	0.02
0.1	2×10^{-3}	0.2
10	0.2	2
1000	20	20

uncompensated resistance. The latter can be minimized by good cell design.

Using cyclic voltammetry to illustrate the influence of charging currents on electrochemical experiments, it is straightforward to compare the magnitudes of the Faradaic and non-Faradaic current densities. In cyclic voltammetry, the potential is changed continuously so that the recorded current always has a component of charging current. The charging current density is given by:

$$j_{charging} = v C_{dl} \tag{6.3}$$

where v is the potential scan rate and C_{dl} the double layer capacitance. Table 6.1 compares values for the peak current density calculated from the Randles–Sevčik equation, Equation (7.53), and the double layer charging current density calculated from Equation (6.3) for a typical set of experimental conditions. It can readily be seen that the double layer charging current density is insignificant at low scan rates but becomes a major factor at high scan rates (due to the different dependences on scan rate: $j_{charging} \propto v$, $j_p \propto v^{1/2}$). The influence of the charging current is also illustrated in Figure 6.3 where the voltammogram at an intermediate scan rate is considered. It can be seen that the double layer charging distorts the whole response. It was noted above that IR_{soln} drop is also a problem at high scan rates. Hence, in practice, cyclic voltammetry is relatively straightforward for many systems with scan rates below $1\ V s^{-1}$ but requires skill and experimental precautions at substantially higher rates.

6.2.3 Electrical Noise

Responses from electrochemical experiments are often affected by electrical noise. Whether this is a problem will depend on the magnitude of the noise compared to the experimental current density. Modern

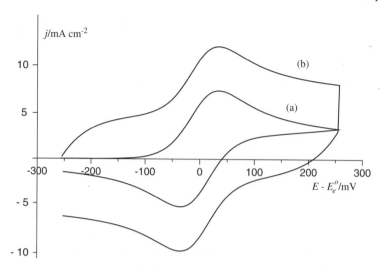

Figure 6.3   Influence of charging current on a cyclic voltammogram recorded with a potential scan rate of $100\,V\,s^{-1}$ for a solution containing 1 mM reactant ($n = 1$ and $D = 10^{-5}\,cm^2\,s^{-1}$). (*a*) Faradaic current only and (*b*) Faradaic + charging current for $C = 50\,\mu F\,cm^{-2}$ and $R_u = 10\,\Omega$.

instrumentation is seldom the source of the noise. Noise most commonly results from the cell or pick-up of signals from other electrical equipment in the locality. Hence, the initial approach should be to examine the cell and ensure that there are no high resistance components (*e.g.* reference electrodes) or poor contacts resulting, for example, from gas bubbles, blocked pores in sinters or corroded crocodile clips. Secondly, consideration should be given to the earthing of all the equipment in the locality to make certain that there is a common earth. Finally, the possibility of placing the cell in a Faraday cage, commonly a small aluminium box, should be examined; this greatly reduces pick-up of electrical signals from the surroundings. A partial move towards this solution is to wrap the cell and contact leads in aluminium foil.

6.2.4   Mass Transport Regime

In the understanding and analysis of data from electrochemical experiments it will be desirable to use the experience (*e.g.* recognition of peak shapes) and equations taken from the literature. To do this, it is essential that the experiment is carried out with a defined mass transport regime and, indeed, the mass transport regime appropriate to the theory. As was discussed in Chapter 1, this will generally be (a) a still solution within a thermostat and containing a large excess of inert electrolyte, *i.e.* conditions that correspond to linear diffusion to a plane electrode, or (b) conditions corresponding to convective-diffusion with a convection regime that may be handled theoretically, most commonly a rotating disc electrode.

6.2.5 Solution Contamination

Interferences from contaminants in solution easily arise in electrochemical experiments. The contaminants may lead to competing electrode reactions or adsorption on the electrode surface, inhibiting or catalysing electrode reactions. Low concentrations of unwanted species can be important especially when, as is common, the electrochemical experiment involves the electroactive species at a relatively low concentration, *ca.* 1–5 mM. Since solutions have frequently been in contact with air, oxygen can be present unless it has been removed with a fast stream of nitrogen or argon and an inert atmosphere above the solution is maintained throughout the experiment; of course, oxygen can be cathodically reduced. Other likely sources of contaminants are the reference electrode (*e.g.* leakage of chloride ion from saturated calomel electrodes) and the inert electrolyte. As the inert electrolyte may be present at a concentration of 1 M, a low % of impurity can result in a concentration of contamination comparable to that for the electroactive species. When working with non-aqueous solvents, there is always water present and, unless very special precautions are taken, its concentration can frequently be > 1 mM.

Another important issue is buffering of the medium. In aqueous systems, many cathode reactions involve the formation of hydroxide while anode reactions often afford protons. Especial care should be taken with electrode reactions in unbuffered 'neutral' solution, since formation of protons or hydroxide at the electrode surface will lead to a boundary layer (where the electron transfer is occurring) that is not neutral and may be of variable pH during the experiment. Certainly, the pH swing within the boundary layer will depend on current density (and therefore, most likely, the reactant concentration) and the mass transport regime. In addition, the pH in the boundary layer may change during an experiment; for example, in voltammetry the pH swing in the boundary layer will increase as the current density increases. Even in acid solutions, the product from some reactions, *e.g.* nitrate reduction, can be determined by the ratio of the reactant and proton concentrations in the solution studied. In the case of nitrate reduction, many products (*e.g.* nitrite, nitrogen dioxide, nitrous oxide, nitric oxide, nitrogen, hydroxylamine and ammonia) can be formed, requiring between $2e^-$ and $8e^-$ and 2–9-H^+/nitrate ion. For example, for the formation of ammonia, the reaction is:

$$NO_3^- + 9H^+ + 8e^- \longrightarrow NH_3 + 3H_2O \qquad (6.4)$$

Hence, unsurprisingly, the product and electrochemical response depend strongly on the ratio of nitrate to protons, and the formation of ammonia clearly necessitates a large excess of acid to avoid a pH change close to the electrode. It is also possible to find conditions where the current is mass transfer controlled with respect to either nitrate or proton.

6.2.6 A Reproducible Electrode Surface

The results from electrochemical experiments can depend strongly on the state of the electrode surface. With solid metal electrodes, the extent of their coverage by oxide or other corrosion films will depend strongly on the history of the surface and results with carbon electrodes can depend on the types of functional groups on the surface and their coverage. Electrode reactions may be catalysed by special sites on an atomic level (*e.g.* kink or step sites) while the rate of corrosion of metals depends strongly on the presence of pits and imperfections in passivating layers. Results can also depend on the surface roughness or dispersion of catalytic materials. Moreover, the influence of surface roughness is not the same for all types of electrode process (Figure 6.4). With reactions that involve adsorption of intermediates (*e.g.* H atoms in hydrogen evolution) or are kinetically controlled, the rate of the electrode reaction will depend on the total surface area exposed to the solution and hence the roughness; the rate of product formation can be greatly enhanced by dispersion of the electrode material in a high area form, *e.g.* Pt in fuel cell electrodes. In contrast, the rate of a mass transport controlled reaction depends only on the apparent geometric area because, with mass transport control, all the species passing through a plane parallel to the surface will undergo reaction. Making the surface rougher will not increase the current density unless the depth of the roughness is thicker than that of the mass transfer layer.

In a previous era, the dropping mercury electrode was extremely popular since a fresh surface was created reproducibly every few seconds. With solid electrodes it is essential to develop a pretreatment procedure that permits the preparation of the same surface for the commencement of each experiment. This may involve (a) polishing the surface with various grades of alumina powder (often starting with 1 µm Al_2O_3 and moving down in stages to 0.05 µm Al_2O_3 powder) on a

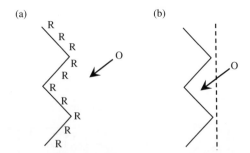

Figure 6.4 (*a*) Kinetically controlled reduction of O – in the example shown, the rate of reaction depends on the number of surface sites available for adsorption of R and therefore on the real surface area; (*b*) mass transfer reduction of O – all O passing through the plane shown will undergo reduction somewhere on the surface and the rate of reduction depends only on the apparent geometric area.

polishing cloth, (b) a chemical treatment such as soaking in acid or alkali to etch away corrosion layers, *etc.*, (c) treatment with a solution of an oxidizing agent to remove adsorbed organic compounds, (d) heating metal electrodes to red heat to burn off organics and (e) an electro-chemical procedure such as cycling the potential between defined limits to oxidize/reduce surface layers and/or adsorbed organics.

There is also a distinction between a reproducible surface and a surface of known composition/structure. Thus the application of a technique to characterize the surface is beneficial. For platinum surfaces, the simplest procedure is to run cyclic voltammograms in 1 M sulfuric acid and to check that the number of peaks and the peak shapes for oxide formation/reduction and hydrogen adsorption/desorption. Figure 6.5 shows a cyclic voltammogram for a 'smooth' polycrystalline Pt wire electrode; clean surfaces give two sharp and clearly distinguished peaks for hydrogen adsorption and three peaks for desorption. The charges associated with these processes (the potential axis is also a time axis, the two quantities inter-related by the potential scan rate) also give a good indication of the surface roughness; for 1e$^-$ oxidation/reduction involving a monolayer, the charge will be $\sim 200\ \mu C\,cm^{-2}$. In the example shown, the actual roughness factor is ~ 3.

6.3 INSTRUMENTATION

Several companies now market reliable instrumentation for electro-chemical experiments and there is no reason to discuss the electronics in this book. In general, the commercial instruments are PC controlled

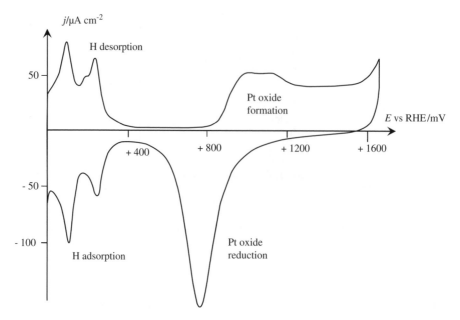

Figure 6.5 Cyclic voltammogram for a clean, polycrystalline Pt wire electrode in 1 M
H$_2$SO$_4$. Potential limits +20 to +1600 mV *vs* RHE. Potential scan rate
100 mV s^{-1}.

and there is software available for the presentation and analysis of
the results on the PC. Problems are much more likely to arise from
the cell or the way the experiment is carried out rather than from the
instrumentation.

The importance of viewing and interpreting qualitatively the results at
an early stage cannot be stressed too strongly; more quantitative use of
the software should be attempted only when the system is understood.
Similarly, there is now excellent software available for simulation of
electrochemical experiments and comparing simulated and experimental
responses can be highly enlightening for both understanding the
mechanism and determining parameters. But, again, the software must
be used correctly. The simulation is based on inputs describing both the
experiment and the chemistry of the system under study and it is also
essential to investigate the sensitivity of the simulated response to
changes in mechanism and numerical parameters. Sensible use of the
software usually requires that the simulation package is understood and
not used as a 'black box'. Moreover, understanding software for
simulation can be a very effective tool for improving understanding of
electrode reactions.

6.4 COMPONENTS IN ELECTROCHEMICAL CELLS

6.4.1 Working Electrode

The experiment must be designed so that the experimental data are totally determined by the chemistry at the working electrode.

A very broad spectrum of electrode materials and forms has been used as working electrodes. With regard to the material of the working electrode there are two types of system: (a) where the choice of working electrode material is predetermined by the nature of the study, *e.g.* the corrosion of nickel or oxygen reduction on Pt electrocatalysts, and (b) where there is a free choice of electrode material, *e.g.* when the working electrode is intended to be inert. In the latter case, the objective will be to select a material that, in the electrolyte medium for the experiment, is stable to corrosion or formation of surface films and does not support competing electrode reactions such as H_2 evolution or O_2 evolution/reduction within the potential range of interest. Common choices are vitreous carbon or gold. The form/structure of the working electrode will be decided by several factors, including the objective of the experiment and the forms of materials available. For example, the structures of a porous gas diffusion electrode for a fuel cell or a metal foam for water treatment are quite different while many experiments demand a flat surface. In addition, carbon is available as graphite, diamond and vitreous carbon as well as carbon, and in many forms, including plates, rods, cloths, felts, foams or powders, while platinum is used as polished metal, a coating on an inert substrate (usually Ti), a metal black and as highly dispersed small centres on an inert substrate such as carbon powder. Other materials may be available only as powder or plates or even as a thin coating on an inert substrate.

For voltammetry and related experiments, a disc (surrounded by an inert polymer sheath, glass or polymer tubing), a wire, a sphere or a small plate is usually employed as the working electrode and the geometric area is usually in the range $0.01–1\ cm^2$. In addition to availability, the choice between the geometries depends on several factors:

1. How is the electrode to be cleaned and prepared for the experiment? For example, a small disc is the easiest to polish while a wire is well suited to being flamed.
2. The mass transport regime; we need to ensure that this is appropriate to the theory being employed to describe the experiment – linear diffusion to a plane electrode or a suitable forced convection regime. While, in practice, equations developed using linear diffusion to a plane electrode are often applicable, we should be

aware that there are always edge effects at, for example, both discs and plates (spades). The assumption in the theory is that diffusion occurs only perpendicular to the electrode surface. This will be true at the centre of an electrode but mass transport can always occur throughout a $90°$ angle at the edge of the electrode (Figure 6.6). Hence, the rate of mass transport to the edge is always high compared to that at the centre. This is confirmed by electro-deposition experiments where the amount of metal at the edge can be seen to be more than at the centre (*e.g.* by using scanning electron microscopy). In voltammetry, however, it is the current density averaged over the whole electrode that is measured and with electrodes with areas greater than a few mm^2 the additional reaction at the edge does not greatly influence the average current density. If the electrode is made much smaller, the relative contribution from the edge to the centre will grow. For example, with a microdisc with a diameter of 10 μm, the edge effect is totally dominant. Such microdisc electrodes cannot therefore be modelled by linear diffusion to a plane electrode and the theory is instead developed using hemispherical diffusion.

3. It is desirable that the same chemistry and the same rate of chemical change occur over the whole electrode surface, otherwise the response is the average response and this is more difficult to interpret. To obtain such uniformity, it is necessary to have a uniform potential and current density over the surface. In Section 6.2.1, it was explained that the instrumentation applies a voltage difference between the working and counter electrode and that the potential difference is given by Equation (6.1). Clearly, the potential will be uniform over the whole surface of the working electrode only if the *IR* drop is the same to all points on this surface. This necessitates that the counter electrode has a geometry that makes it equidistant from all points on the surface of the working electrode. Figure 6.7(a)–(d) shows some acceptable cell geometries. An

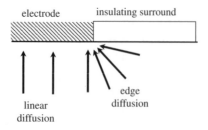

Figure 6.6   Illustration of the 'edge effect' on diffusion to a real electrode.

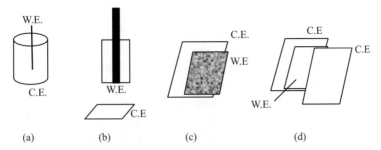

Figure 6.7 Examples of good cell geometry: (*a*) wire working electrode (W.E.) surrounded by cylindrical counter electrode (C.E.), (*b*) disc working electrode facing a larger plate counter electrode, (*c*) a plate working electrode with an insulated back face with a larger parallel plate counter electrode and (*d*) a plate working electrode with larger parallel plate counter electrodes equidistant on either side.

example of poor design would be a cell with a plate working electrode with both sides active and a counter electrode facing only one surface. The length of the solution path to the front and back of the working electrode will be quite different and the driving force for electron transfer cannot be the same on the two sides of the electrode. The current density on the back will be lower than the face opposite the counter electrode. Reflecting the discussion in Section 6.2.1, the likelihood of a significantly uneven potential distribution will increase with increasing current density and solution resistance. For low current experiments in highly conducting solutions, the variation in potential over the surface will be smaller and the requirement for good cell geometry becomes more relaxed.

6.4.2 Counter Electrode

The job of the counter electrode (also called the secondary or subsidiary electrode) is to provide a current of equal magnitude but opposite sign to that at the working electrode (if the reaction at the working electrode is an oxidation, reduction will occur at the counter electrode and *vice versa*) without interfering with the response of the working electrode. In fact, the electrode reaction at the counter electrode will be that which is most facile in the electrolyte solution – if an oxidation, the reaction that occurs at the least positive potential or if a reduction that which takes place at the least negative potential. The charge passed at the counter electrode will always be the same as the charge at the working electrode. Hence, the two key requirements for the counter electrode are an electron-transfer reaction that occurs readily and whose chemistry does not interfere with the response at the working electrode. The latter will be most important in long timescale experiments. Notably, even H_2 and O_2 evolution lead to a local change in pH (Table 1.2), and this can change the bulk pH over an extended period. In an undivided cell, the overall chemical change in the electrolyte solution is the sum of the working and counter electrode reactions; in these circumstances, the appropriate choice of counter electrode reaction can be used positively to maintain the pH constant or to replenish other electrolyte components (*e.g.* in a copper plating bath the counter electrode reaction could be copper dissolution).

In non-aqueous solvents, the counter electrode chemistry can be more of a problem. In particular, aprotic solvents undergo complex decomposition reactions, either directly in an electron-transfer reaction or as a result of reactions of acids/bases formed in the electrode reactions. Hence, there is more reason to introduce a separator between the working and counter electrodes for all but short timescale experiments. This, of course, assumes the availability of a suitable separator, see below, and also increases the resistance of the cell. An alternative approach is to use an undivided cell and to find a benign counter electrode reaction. In cathodic electrosynthesis, one such approach is to use a dissolving metal anode, either Al, Mg or Zn, although the anode metal becomes a stoichiometric reagent rather than simply an electrode.

In the discussion of the working electrode, it was pointed out that the design of the counter electrode contributes to an appropriate cell geometry. Also, it is sensible cell design to make the counter electrode as large as possible since this reduces the current density drawn at this electrode. With regard to the material for the counter electrode, a wide choice is often available. When the counter electrode reaction is an oxidation, one needs to be cautious of low rate corrosion of the anode in long timescale experiments; there are, for example, reports of the transfer of Pt from counter to working electrode influencing experimental data. Carbon is often a safe choice of anode counter electrode. If the counter electrode reaction is hydrogen evolution, nickel or steel is usually a safe alternative to the widely used but expensive platinum.

In a two-electrode cell, one electrode has to combine the functions of the reference and counter electrodes.

6.4.3 Reference Electrode

The reference electrode should maintain a potential that is constant throughout the experiment and have the same value every day. In particular, it should be independent of current density. This essentially means that the couple employed should have rapid kinetics; a plot of *j vs E* close to the equilibrium potential should be very steep [see Figure 1.6 and Equation (1.48) and the associated discussion]. The requirement is easier to meet in a three-electrode cell because the potentiostat is designed to ensure that the current through the reference electrode is very low indeed. In a two-electrode cell, the reference electrode must pass the full cell current and this will usually be much higher. It is also advantageous that the chemistry of the reference electrode is defined and that the reaction follows the Nernst equation with a known dependence of the potential on the concentration of species in solution.

Table 6.2 lists some generally used aqueous reference electrodes. By convention, the potential of the standard hydrogen electrode (SHE) is zero. This electrode consists of a high area Pt electrode (usually a Pt black) in a solution containing 1 M proton and saturated with hydrogen (usually by passing H_2 through the solution or trapping a H_2 atmosphere above the solution). The reversible hydrogen electrode (RHE) differs only in that the electrolyte is the same as that under study (not 1 M acid); as a result, its potential will depend on the pH of this solution. The three mercury-based electrodes are also convenient and each consists of mercury in contact with an insoluble mercurous salt; their potential is a function of the concentration of the anion present in the electrolyte. This can be seen by writing the Nernst equation for, for example, the chemistry of the saturated calomel electrode (noting that

Table 6.2 Some common reference electrode in aqueous solutions and their potentials.

Name	Abbreviation	Electrolyte	Reaction	Potential vs SHE/mV
Standard hydrogen electrode	SHE	1 M H^+	$2H^+ + 2e^- \rightleftarrows H_2$	0
Reversible hydrogen electrode	RHE	Solution under study	$2H^+ + 2e^- \rightleftarrows H_2$	-60pH
Saturated calomel electrode	SCE	Saturated KCl	$Hg_2Cl_2 + 2e^- \rightleftarrows 2Hg + 2Cl^-$	$+241$
Mercury/ mercurous sulfate	–	Saturated K_2SO_4	$Hg_2SO_4 + 2e^- \rightleftarrows 2Hg + SO_4^{2-}$	$+640$
Mercury/ mercurous oxide	–	1 M NaOH	$HgO + H_2O + 2e^- \rightleftarrows Hg + 2OH^-$	$+926$
Silver/silver chloride	–	Saturated KCl	$AgCl + e^- \rightleftarrows Ag + Cl^-$	$+197$

the activity of elemental mercury is one) and then recognizing that the concentration of mercurous chloride in solution will be a constant because the solution is always saturated with this species:

$$E_e = E_e^0 + \frac{2.3RT}{2F} \log \frac{c_{Hg_2Cl_2}}{c_{Cl^-}^2}$$

$$= E_e^{o'} - \frac{2.3RT}{F} \log c_{Cl^-}$$

(6.5)

Whenever possible, to avoid contamination of the working electrode solution by additional anions, the reference electrode is chosen so that the anion of the mercury reference electrode is the same as that in the solution under study. The silver/silver chloride reference electrode is another example of a metal in contact with an insoluble salt and is the reference electrode of choice for studies at elevated temperatures. Hydrogen reference electrodes can be inconvenient as they require hydrogen gas; consequently, the other four electrodes in the table are the most common general experimental reference electrodes. Other couples are convenient for particular studies. For example, in the deposition of a metal, a convenient reference electrode can be simply the metal itself dipping into the bath being employed; with such a M/M^{n+} reference electrode, the overpotential is the same as the measured potential since the reference point is the equilibrium potential of the M/M^{n+} couple.

Conversion of potential scales between different reference electrodes should be simple. It requires only the addition or subtraction of the difference in the potentials of the two reference electrodes. Figure 6.8 illustrates a procedure to ensure the correct transfer between potential scales. In this example, a peak is observed at $+290\,mV$ *vs* SCE. By placing a marker on the scale, it is apparent that this is also $(290 + 241) = 531\,mV$ *vs* SHE and $(531 - 640) = -109\,mV$ *vs* $Hg/HgSO_4$.

Even now, much less is known about reference electrodes in non-aqueous solvents. Numerous non-aqueous media have been used for electrochemical studies and precise measurements in each require the development of a new, reliable and reproducible reference electrode. Moreover, attempts to develop such reference electrodes have not always been successful. In non-aqueous media, the potentials of reference electrodes commonly drift with time and do not take up the same potential every day or show the same behaviour in different laboratories. Hence, the reference electrode in non-aqueous studies are often 'make do' choices and may be reproducible only to ± 10–$100\,mV$. In such systems, it can be sensible to use an internal standard, *i.e.* a species

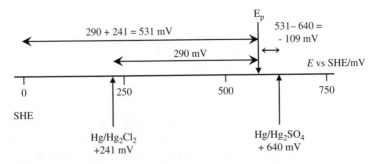

Figure 6.8 A simple procedure to ensure correct transfer between potential scales.

with a simple, reversible electron-transfer reaction, either present during the study or added for a final experiment at the end of a series of experiments. This couple will take up the same potential on all occasions and will act as a reliable reference point for the potential scale. In organic solvents, a common choice for the internal standard is the Cp_2Fe^+/Cp_2Fe couple.

There are no ways to compare precisely potentials in different solvents. This is because at the junction between two solvents there is a liquid junction potential that cannot be calculated or measured. All estimates of the liquid junction potential require an assumption that cannot be fully substantiated. A way to get an approximate comparison of potential in different solvents is again to use the Cp_2Fe^+/Cp_2Fe couple:

$$Cp_2Fe^+ + e^- \rightleftharpoons Cp_2Fe \tag{6.6}$$

The assumption that this reaction has the same formal potential in all solvents is the equivalent of assuming that the solvation of both the ferrocinium cation and the ferrocene molecule are the same in all solvents. The error introduced by this assumption cannot be calculated but the couple is selected because the cyclopentadienyl ligands closely surround the Fe^{III} and Fe^{II} and therefore largely protect them from interaction with the solvent.

6.4.4 Electrolyte Solution

In addition to the species particular to the study (electroactive species, additives, *etc.*) the solution for most electrochemical experiments contains a large excess of an inert electrolyte (also called the base electrolyte). This has several roles:

- To give the solution ionic conductivity and thereby lower its resistance. Remember, electron transfer at the electrode surfaces can occur only if there is a mechanism for the charge to pass through the solution between the electrodes (see Figure 1.1 and the associated discussion). In addition, to avoid unwanted effects on the measured potentials from IR drop, the resistivity of the solution should be as low as possible. As noted in Chapter 2, there are large differences in the conductivity of aqueous and non-aqueous solvents and even between aqueous solutions of acids, bases and neutral salts.
- To avoid migration having a significant role in the mass transport of the reactant and/or the product of electron transfer. Migration is the mechanism by which charge is moved through solution. It results from electrostatic forces and the charge is carried by all ions in solution and the fraction of the charge carried by each ion depends strongly on their relative concentrations. With a 100–1000-fold excess of inert electrolyte, effectively all of the charge is carried by the ions from the electrolyte.
- To provide ions to form the double layer at the electrode surface. With low ionic strength solutions the double layer stretches out into solution and this has the effect of smearing out the potential field at the electrode/solution interface and slowing down electron transfer. Hence a high ionic strength facilitates simple electron transfer kinetics.
- The concentrations of ions of the inert electrolyte are the major contributors to the ionic strength of the medium. In consequence, the activity coefficients of all ionic species in solution, calculated, for example, using Equation (2.7), are similar. For couples such as $M^{n+}/M^{(n-1)+}$, the activity coefficients in equations such as the Nernst equation will therefore cancel, removing the need to consider non-ideal behaviour.
- When an acid or base, the electrolyte acts as a buffer preventing a pH swing within the reaction layer at the electrode surface when the electron-transfer reaction leads to the consumption or formation of H^+ or OH^-. When the pH of the electrolyte is in the pH range 2–12, the advantages and consequences of adding a buffer (*e.g.* phosphate or borate) should be considered.

6.4.5 Separators and Membranes

In many situations, it is desirable to have a separator between the working electrode and reference electrode and/or the working and counter electrode although, of course, electrical contact must be maintained. The most common separator materials are glass sinters, porous ceramics, porous polymers or ion-permeable membranes. The latter are thin films of charged organic polymers, such as Nafion^{TM}, designed to transport selectively either cations or anions (Section 2.4). Such membranes are much more selective than porous materials in the transport that is allowed but all separators allow the transport of solvent and neutral molecules at a finite, if relatively low, rate while porous materials will show no selectivity in the ionic transport and may allow quite facile transport of neutral species. Ion-permeable membranes are designed for operation in aqueous systems and usually perform poorly in non-aqueous systems.

In three-electrode cells, the instrumentation ensures an almost zero current through the reference electrode and hence the resistance of the separator between reference and working electrode is not a major issue. Commonly, the separators are ceramics with a rather low porosity. Between the working and counter electrodes two issues become important particularly in longer timescale experiments where substantial chemical change is occurring, *e.g.* an electrolysis cell for synthesis: (a) the resistance should be low and (b) since the composition of both anolyte and catholyte will change substantially (as determined by Faraday's law) it is desirable to prevent transport of all but ions through the separator in both directions – it is undesirable to lose or gain materials from/to the working electrode compartment. While glass sinters can be used in the laboratory, their limited powers of separation need to be recognized. In electro-chemical technology the separator is invariably an ion-permeable membrane although it is still necessary to operate them in conditions where their deficiencies do not degrade process performance unreasonably.

Always, the separator between the working and counter electrodes should have an area larger than that of the electrodes so that a uniform current path between the electrodes is possible. Otherwise, the current distribution over the working electrode will not be uniform.

6.5 SOME CELL DESIGNS

At this point we must distinguish three types of experiment and there-fore three types of cell. In the first, the objective is to study mechanism and kinetics and it will be advantageous to carry out many experiments without significant changes to the system. Hence, the cell will be designed with a small working electrode so that the charge passed during an experiment is low. The cell design must ensure an appropriate mass transport regime while it may also be important that the working electrode may be cleaned. In the second, the objective is to bring about chemical change, *e.g.* a synthesis cell, a water treatment cell or a fuel cell. Here, a rapid electrolysis will require larger electrode areas or more specifically a high ratio of electrode area to electrolyte volume. In addition, while a high rate of mass transport is desirable it is not necessary that the mass transfer regime is describable. In the third type of experiment, the shape of the working electrode is predetermined and the cell design must facilitate this electrode shape. For example, a cell for electroplating a complex shaped object, *e.g.* a teapot or a tool. Only the first two types of cell will be considered further here.

Figure 6.9 shows an example of a two-electrode cell suitable for voltammetry under conditions of linear diffusion to a plane electrode. It is essentially a covered beaker with a gas inlet for removing oxygen. Such cells are readily built, for example, based on a three-necked, round bottomed flask or other laboratory glassware but are suitable only for experiments with low currents. Figures 6.10 and 6.11 show two examples of three-electrode cells, again intended for voltammetry and related experiments. The first is designed with a wire working electrode for still solutions while the second has a rotating disc working electrode. Both have the reference electrode mounted in a Luggin capillary whose tip is ideally positioned ~ 1 mm from the surface of the working electrode. In addition, the geometry of the counter electrodes provides approximations to the ideal cell geometry to give a uniform current distribution. While these cells require a glass blower for their fabrication, they will give much better results when the current density is higher. With additional effort, a glassblower can also put a water jacket around the cell to

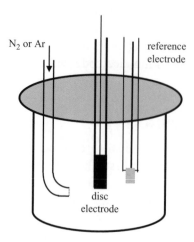

Figure 6.9 Two-electrode, single-compartment cell with a disc electrode.

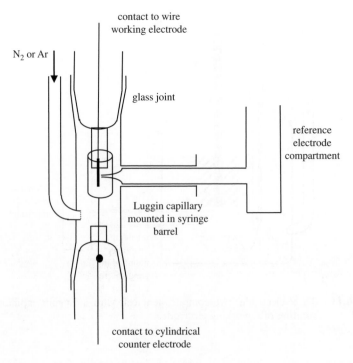

contact to wire
working electrode

N_2 or Ar

glass joint

reference
electrode
compartment

Luggin capillary
mounted in syringe
barrel

contact to cylindrical
counter electrode

Figure 6.10 Three-electrode, two-compartment cell with a Luggin capillary and a wire working electrode.

allow circulation of water from a thermostat and therefore permit work at a constant, known temperature.

Figures 6.12 and 6.13 are cells intended for synthesis. Hence, they have large electrodes and relatively small volumes of solution. Both have separators to avoid the back reaction at the counter electrode and mixing of the chemistries at the two electrodes. An important feature is that the separators have an area as least as big as the electrodes. Also, since the concentration of reactant is likely to be high and efficient mass transport conditions will be employed, the cell current may be high:

$$I = nFAk_mc \qquad (6.7)$$

In consequence, the potential and/or current distribution will be critical and, therefore, the working and counter electrodes are parallel to each other and also as close as possible to reduce cell resistance. The cell in Figure 6.12 is again made from glass and has a Luggin capillary to allow controlled potential electrolyses. The working and counter

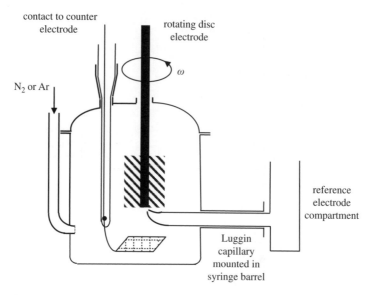

Figure 6.11   Three-electrode, two-compartment cell with a Luggin capillary and a rotating disc working electrode.

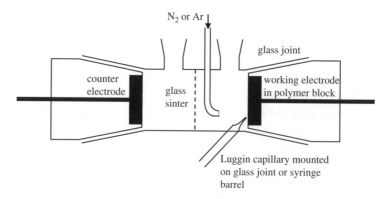

Figure 6.12   Glass cell for controlled potential electrolysis.

electrodes are large discs mounted in PTFE shrouds although similar cells with two metal mesh electrodes are also common. Figure 6.13 shows a small flow cell that will ape the performance of larger parallel plate reactors. Here the separator is an ion-exchange membrane and the cell is held together with a clamp. The electrolyte flow channels are formed from polymer sheets that also act as gaskets to avoid solution leakage so that there may only be a ~0.2 cm gap between electrode and

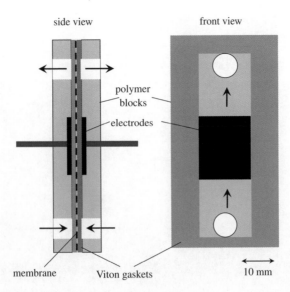

side view   front view

polymer
blocks

electrodes

membrane   Viton gaskets   10 mm

Figure 6.13 Construction of a small, laboratory flow cell with a membrane. The arrows indicate the electrolyte flows.

membrane. In such cells it is difficult to introduce a Luggin capillary (although a thin polymer tube can mimic a Luggin capillary quite well) so they are usually operated at constant current. Figure 6.14 is a photograph of a small flow cell and flow circuit. The cell has electrodes 2×1 cm and can make ~ 15 mmol of product per hour.

6.6 WHAT IS CONTROLLED?

In the literature, a very clear distinction is often made between controlled potential and controlled current experiments and, for example, claims are made that controlled potential electrolysis leads to a more selective reaction. It should, however, be remembered that, for any solution, current density and potential are not independent parameters but are related by the voltammogram for the solution. A particular example is helpful.

Example: An electrolysis for the oxidation of R to O is carried out in a divided cell with a large ratio of anode area to anolyte volume and the anolyte is well stirred for rapid electrolysis. Periodically during the electrolysis a voltammogram is recorded at an inert rotating disc electrode. During a controlled potential electrolysis, both current and charge are monitored as a function of time;

Figure 6.14 A small flow cell with flow circuit, including reservoir and peristaltic pump.

during the controlled current electrolysis, the potential is monitored as a function of time. It will be assumed that that the product O is electroinactive and the wave for the oxidation of R is well negative to the potential where O_2 is evolved.

The potential for a controlled potential electrolysis is, whenever possible, selected to be in the mass transfer controlled plateau since this represents the maximum rate for the conversion. Figure 6.15 summarizes the results. Figure 6.15(a) shows the way in which the voltammograms change with charge passed (also time) during the electrolysis. The wave for the oxidation of R continues to be observed but its limiting current decays with time as the reactant is consumed. In fact, since the reaction is first order in R and the electrolysis is under

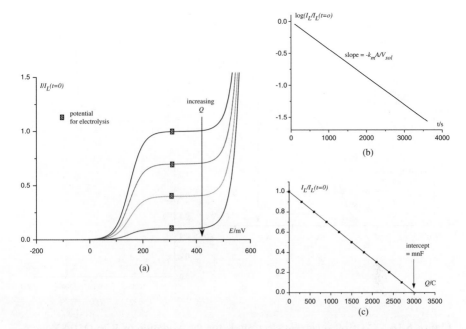

Figure 6.15 Controlled potential electrolysis for the oxidation of R to O. (*a*) Voltammograms at a rotating disc electrode recorded at intervals during the electrolysis; plots of log $I_L/I_L(t=0)$ *vs* (*b*) t and (*c*) Q.

mass transfer control, it can be shown that:

$$c_R(t) = c_R \exp -\left(\frac{k_m A t}{V_{sol}}\right) \tag{6.8}$$

where k_m is the mass transfer coefficient, A the electrode area and V_{sol} the volume of solution being electrolysed. Since the limiting current from the voltammogram is proportional to $c_R(t)$, a plot of log $I_L/I_L(t=0)$ *vs* t is linear, see Figure 6.15(b); the mass transfer coefficient for the cell conditions may be estimated from the slope. Figure 6.15(c) shows another useful plot, namely, $I_L/I_L(t=0)$ *vs* Q, the charge passed; this is also linear and n can be calculated from the intercept on the Q axis. In each of these plots, the limiting current on the RDE could be replaced by the cell current [also proportional to $c_R(t)$] or the concentration determined in any other way. At the end of the electrolysis an appropriate analytical procedure (spectroscopy or chromatography) could be used to confirm that R is the (sole) product or to determine the selectivity of the reaction.

Usually, the current density for a constant current electrolysis is chosen to be towards the foot of the wave. Figure 6.16(a) shows the

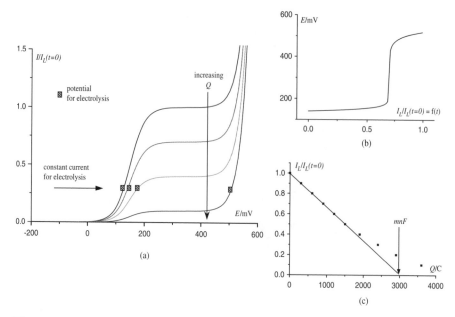

Figure 6.16 Controlled current electrolysis for the oxidation of R to O. (*a*) Voltammograms at a rotating disc electrode recorded at intervals during the
electrolysis; plots of (*b*) *E vs* $I_L/I_L(t=0)$, proportional to the electrolysis
time (*t*), and (*c*) $I_L/I_L(t=0)$ *vs Q*.

voltammograms as the electrolysis is carried out. Again the limiting
current diminishes as a linear function of the charge passed; $Q\ (=It)$
and *n* can be calculated from the intercept on the *Q* (or *t*) axis. It can be
seen from the voltammograms and also Figure 6.16(b) that during the
early part of the electrolysis the potential varies little and the selectivity
of the reaction will be the same as for the controlled potential
electrolysis. Indeed, it is not until $I > I_L(t)$ that there is a large shift in
potential to a value where another oxidation can occur. If the next
easiest oxidation is oxygen evolution, there may be no loss in selectivity
for the conversion of R into O, although there will certainly be a loss in
current efficiency. Figure 6.16(c) shows the plot of I_L *vs Q* for the constant current electrolysis. Early on in the electrolysis, this plot is again
linear and the line can be extrapolated to $I_L/I_L(t=0)=0$ to allow the
determination of *n*. Once a competing oxidation commences, the data no
longer falls on the line as a fraction (increasing with time) of the charge
is consumed in the competing reaction.

In the laboratory and on a larger scale, the attraction of constant
current electrolysis is the simpler and cheaper instrumentation (the
difference in cost escalates as the cell current increases). The drawback is

that the rate of conversion is slower if a high selectivity and current efficiency is to be guaranteed; to stay close to the initial potential for most of the electrolysis then the cell current must be much less than the limiting current.

Further Reading

1. Southampton Electrochemistry Group, *Instrumental Methods in Electrochemistry*, Ellis Horwood, Chichester, republished 2001.
2. *Laboratory Techniques in Electroanalytical Chemistry*, ed. P. T. Kissinger and W. R. Heineman, Marcel Dekker, New York, 1996.
3. T. A. Davis, J. D. Genders and D. Pletcher, *A First Course in Ion Permeable Membranes*, The Electrochemical Consultancy, Romsey, 1997.
4. F. C. Walsh, *A First Course in Electrochemical Engineering*, The Electrochemical Consultancy, Romsey, 1993.
5. D. Pletcher and F. C. Walsh, *Industrial Electrochemistry*, Chapman & Hall, London, 1990.

CHAPTER 7

Techniques for the Study of Electrode Reactions

7.1 INTRODUCTION

This chapter discusses experimental techniques for the definition of the mechanism and kinetics of electrode reactions. In Chapter 1, it was emphasized that even the simplest redox reaction is a multistep process involving both mass transport and electron transfer. Complete characterization of such a reaction would require determination of:

- the formal potential, E_e^o, for the couple;
- the standard rate constant, k_s (or the exchange current density, j_o) and the transfer coefficient, α, for the couple;
- the diffusion coefficients for the oxidized and reduced species.

Hence, as each electroanalytical technique is introduced, one objective will be the recognition of a simple redox process and understanding how these parameters can be determined. Perhaps, it is also desirable to determine how the reaction thermodynamics and kinetics change with experimental conditions such as the solution composition and temperature. It was, however, also noted in Chapter 1 that most electrode reactions are more complicated, with the possible involvement of adsorption, coupled chemical reactions, phase formation and multiple electron transfers. Hence, the second objective will be to develop tests that allow us to understand which complications are important for the system under study and then to examine finer details of the mechanism.

A First Course in Electrode Processes, 2nd Edition
By Derek Pletcher
Published by the Royal Society of Chemistry, www.rsc.org

With complex multistep reactions it has to be recognized that it will not be possible to gain information about all steps; inevitably, the focus will be on the rate-determining step. In the development of electrochemical technology, the challenge is usually to increase the current density and/or selectivity of the electrode reaction and this can be achieved only through the identification of the rate-determining step; the factors influencing the rate of electron transfer, mass transport, *etc.* are different and hence the approach to achieving the target depends totally on knowing the rate-determining step. In all systems, it is critical that the products are fully defined and that the electron stoichiometry is known (as well as how these depend on the electrode potential, solution composition, *etc.*) before detailed mechanistic studies are undertaken.

In this chapter, both steady state and non-steady state techniques will be considered and it will be stressed that, whenever possible, experiments should be carried out where the conditions are fully controlled. For example, the mass transfer regime should be controlled (as well as describable by mathematical equations) and experiments should be carried out in a thermostat. Steady state techniques will give quantitative information only under conditions where there is a single rate-determining step, *i.e.* when one step is clearly slower than all other steps. It is, however, possible to vary the electrode potential and/or the mass transport regime to enforce rate control by a single step. In addition, under conditions where there is mixed control by mass transport and another process (most commonly electron transfer although a chemical reaction is also possible), there is a general approach to obtaining pure kinetic information. This involves (a) the measurement of the current density as a function of a parameter that determines the rate of mass transport (*e.g.* the rotation rate of a disc electrode, ω); (b) extrapolation of the data to a hypothetical situation where the rate of mass transport is infinite (*e.g.* where $\omega \to \infty$) when the competing step must become rate determining. The latter is most conveniently achieved using a linear plot derived from theory (see later) and this is why the mass transfer regime must be described accurately by a set of equations that can be solved. A related concept underlies the application of non-steady state techniques such as cyclic voltammetry, chronoamperometry and impedance. In these techniques, potential scan rate, time and frequency, respectively, are used to vary the rate of non-steady state diffusion. These concepts will be illustrated many times later in this chapter.

It is important to recognize the procedure used to extract mechanistic and kinetic information from the data obtained during electrochemical experiments. Invariably, the procedure involves comparison of the experimental data with a response predicted from a theoretical description of the experiment. As discussed around Figure 1.7, the latter may be a qualitative discussion leading to the general form of the response or a rigorous mathematical description of the experiment or a simulation leading to an exact prediction of the response. The prediction may take the form of the whole response or equations related to particular features of the responses (*e.g.* peak potentials, peak current densities in cyclic voltammetry). These days, when experimental data are available in digital form, comparison of the whole response with the theoretical or simulated response is to be encouraged. Again, both qualitative and quantitative approaches will be illustrated repeatedly later in this chapter. In both cases, it is essential to have a mode of mass transport that is well defined; hence the predominance of experiments carried out under conditions of (a) linear diffusion to a plane electrode, (b) hemispherical diffusion to a microdisc electrode or (c) a simple form of convective diffusion such as a rotating disc electrode. Fortunately, most of the theory required for quantitative analysis is available in the literature but it is still incumbent upon the experimentalist to ensure that the mass transport regime and other conditions in their experiment are consistent with those assumed in the theoretical description of the experiment.

Throughout this chapter, the techniques are discussed in terms of the study of electrode reactions. It is important to recognize, however, that a 'coupled chemical reaction' can also be regarded as a chemical reaction where the electrode reaction is generating one reactant. Hence, the same experiments provide a way to expand knowledge of the chemistry of reactive intermediates. For example, much of our knowledge of ion radical chemistry results from electrochemical experiments. The presentation in this chapter is intended to avoid extensive mathematics. Frequently, equations will be presented without derivation. However, it is hoped that the origins of the equations will be clear and the detailed mathematics are available in the texts at the end of the chapter as well as in the original literature.

7.2 STEADY STATE TECHNIQUES

7.2.1 Electrolysis/Coulometry

It makes no sense to study a reaction in any detail when the stoichiometry and products are not fully defined. Early in the investigation of a new system, it is to be recommended that:

1. The chemical stability of the reactant under the electrolysis conditions should be confirmed. This can readily be achieved using a

spectroscopic or chromatographic method at intervals after the reactant has been dissolved in the electrolyte.

2. The stoichiometry with respect to electrons (*i.e.* the number of electrons per molecule of reactant transferred in the electrode reaction, n) be found.

3. The major products and their yields should be determined. If appropriate, the electrolysis conditions should be modified to optimize the yield of a desired product. Both the current efficiency and the chemical yield are of concern. The current efficiency is the yield based on the charge passed (competing electrode reactions will lead to a current efficiency <1.0, see equation (1.24)). The chemical yield is based on the reactant consumed, (do the reactive intermediates formed at the electrode have competing pathways?). The determination of the current efficiency and/or chemical yield will require analysis of the electrolyte solution at the end of the electrolysis (usually using spectroscopy or chromatography) and perhaps work up of the electrolysis to isolate the products. Clearly, such procedures are most easily achieved with concentrated solutions of reactant.

4. The chemical stability of the reactant and major products in the electrolyte solution towards the end of the electrolysis be confirmed. Remember that electrode reactions commonly lead to substantial pH changes of the electrolyte – base is often generated at the cathode and acid at the anode – and the stability of compounds can vary significantly with pH.

This information can be determined only by experiments where the reactants are fully converted into products, usually by a controlled potential electrolysis. It is convenient and desirable that the chemical change is completed rapidly (*ca.* <30 min). This may be achieved by designing a cell with a large ratio of electrode area, A, to volume of electrolyte, V_{sol}, as well as efficient mass transport to give a high value of the mass transfer coefficient, k_m. In fact, the conversion is given by:

$$\frac{c(t)}{c_{t=0}} = \exp -\left(\frac{k_m A t}{V_{sol}}\right) \tag{7.1}$$

Such cells were discussed in Section 6.6. In contrast to most electrochemical experiments, it is not necessary that the mass transfer regime can be described by mathematical equations; it is necessary only that it is constant throughout the experiment. Thus, a magnetic stirrer or a flow of gas bubbles may be used.

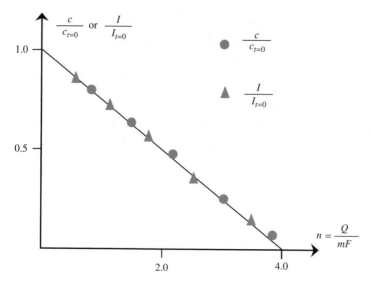

Figure 7.1 Determination of the number of electrons/molecule involved in the oxidation of ethanol at a Ni anode in 1 M NaOH; dimensionless plots of the concentration of ethanol and current *vs* charge passed. The data show that $n = 4$, and chromatographic analysis would show that acetate is formed in almost quantitative yield.

The *n*-value will be established by application of Faraday's law:

$$Q = \int I \mathrm{d}t = mnF \tag{7.2}$$

In one calculation, Q could be the charge for the complete conversion of reactant into products (as indicated by the electrolysis current dropping to zero or, at least, the level of the background current) and m the number of moles of reactant added to the electrolyte. It is, however, more informative to monitor the current and the concentration of reactant (using, for example, spectroscopy or chromatography) as a function of charge passed. Plots of both $I/I_{t=0}$ *vs* Q/mF and $c/c_{t=0}$ *vs* Q/mF should fall on the same line and this will extrapolate to an intercept on the Q/mF axis equal to n (Figure 7.1). If these plots are not linear, the system is more complex; it could indicate, for example, that the products are not completely stable in the electrolysis medium and slowly convert into another species that is electroactive.

In systems with coupled chemistry it is sometimes found that the *n*-value determined by coulometry differs from that estimated by electroanalytical techniques such as cyclic voltammetry. This difference arises from differences in the timescales of the experiment. During a complete electrolysis taking *ca.* 30 min, homogeneous chemistry is more likely than during the few seconds for an electroanalytical experiment.

7.2.2 Steady State Current Density *vs* Potential

Historically, the collection and analysis of steady state *j vs E* data have contributed much to our understanding of electrode reactions and, even now, it offers a useful approach to investigating complex reactions (Chapter 5) and also corrosion. Originally, the current density was varied manually at intervals and, at each current density, the potential was allowed to relax to a steady value. With the advent of potentiostats, it became more common to vary the potential manually or to use a slow potential scan. The data can be interpreted using both *j vs E* and log *j vs E* plots. It is important that Tafel analysis is attempted only when the *j vs E* data can be obtained over a wide potential range leading to several orders of magnitude of current density. Again, exact description of the mass transport regime is not necessary in such experiments but the use of a constant regime is to be recommended. Indeed, a mass transport regime that can be varied (*e.g.* using a rotating disc electrode) has the advantage that the importance of mass transfer in determining the current density can be immediately established.

Practically, there can be a problem in the concept of 'the steady state'. In theory, the current density should reach a steady value within 60 s; in potential sweep experiments, the definition of steady state would be that the response is independent of scan rate over a range of values. In practical systems, however, the electrode surface may change with time; the surface may poison for an electrocatalytic reaction, an impurity metal in the electrolyte may deposit, a corrosion film may form or the surface area of the electrode may change (due to deposition or active corrosion). Such changes may be unavoidable under the conditions of interest and the current density may change (usually, but not necessarily, decrease) for very long periods of time. The experimentalist then must decide which surface is of interest and the timescale when the current density best corresponds to the steady state value for this surface.

7.2.3 Rotating Disc Electrodes (RDEs)

The rotating disc electrode was introduced in Sections 1.4.2 and 1.5 as an example of an experiment where mass transport is by convective diffusion. It was stated that the RDE gives a well-defined, steady state mass transport regime and that the mass transfer coefficient is a function of the square root of the rotation rate of the disc. The concept of the Nernst diffusion layer provides a basis for a good qualitative understanding of experiments and more quantitative models allow an exact prediction of the voltammetric response.

A RDE consists of a polished disc of the electrode material, radius 0.1–1.0 cm, surrounded by an insulating sheath of significantly larger diameter; only the front face of the disc is exposed to the electrolyte solution and the surfaces of the disc and sheath need to be polished so that there is no step where the disc and sheath meet. The structure is rotated about an axis perpendicular to the surface of the disc. The mass transfer regime resulting from rotation of the disc structure can easily be seen by dropping a crystal of $KMnO_4$ or a dye below the disc in a beaker of water. The RDE acts as a pump and solution is pulled vertically upwards towards the disc and then is thrown outwards because of the presence of the solid surface. The rate of pumping can also be seen to increase with the rotation rate of the disc, ω. Figure 7.2 depicts the qualitative motions of the solution and sets out the cylindrical polar coordinates usually used for this geometry (Figure 7.2c):

x, distance perpendicular to the disc surface;
r, distance from the centre of the disc in the plane of the disc;
θ, the angle around the disc.

A detailed mathematical description of the hydrodynamics of a vessel containing a RDE is beyond the scope of this book (the interested reader should refer to V.G. Levich, *Physicochemical Hydrodynamics*, Prentice Hall, Englewood Cliffs, NJ, 1962) but the important conclusions are set out in dimensionless form in Figure 7.3.

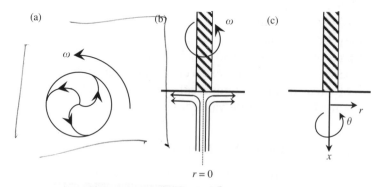

Figure 7.2 Flow patterns close to the rotating disc surface: (*a*) view from below, (*b*) view from the side and (*c*) cylindrical polar coordinates for the quantitative discussion of the hydrodynamics of the RDE.

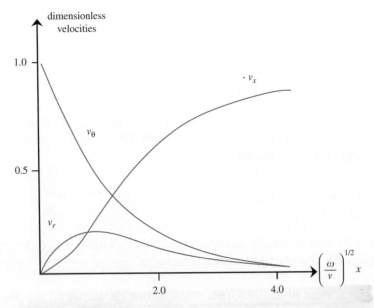

Figure 7.3 Dimensionless representation of the variation of the solution velocities in the x, r and θ directions as a function of distance from the RDE surface; ω is the rotation rate of the disc (in radians s^{-1}) and ν is the kinematic viscosity of the solution (*i.e.* viscosity/density).

The following should be noted:

1. In the bulk solution, the velocity of solution movement towards the disc, v_x, is constant, but as it approaches the solid surface it decelerates and at the surface, $x=0$, it must become zero. Levich demonstrated that close to the surface the velocity is given by:

$$v_x = \frac{-0.51\omega^{3/2}}{\nu^{1/2}}x^2 \tag{7.3}$$

2. The solution is dragged around by the solid surface and hence the rotational movement of the solution is strongest at $x=0$ and drops off smoothly with distance from the surface.
3. The centrifugal motion is zero at the surface, passes through a broad maximum within the boundary layer and then drops off smoothly with x.

It is always possible to write an expression for the variation in concentration resulting from convective diffusion. In general, it will need to be in three dimensions and contain terms that describe diffusion (Fick's laws) and convection. In Cartesian coordinates this equation will have the form (assuming the electroactive species is stable in the electrolyte):

$$\frac{\partial c}{\partial t} = D\left(\frac{\partial^2 c}{\partial x^2} + \frac{\partial^2 c}{\partial y^2} + \frac{\partial^2 c}{\partial z^2}\right) + \left(v_x \frac{\partial c}{\partial x} + v_y \frac{\partial c}{\partial y} + v_z \frac{\partial c}{\partial z}\right) \quad (7.4)$$

where the first and second terms on the right-hand side describe diffusion and convection, respectively; or in the cylindrical polar coordinates appropriate to the RDE:

$$\frac{\partial c}{\partial t} = D\left(\frac{\partial^2 c}{\partial x^2} + \frac{\partial^2 c}{\partial r^2} + \frac{1}{r}\frac{\partial c}{\partial r} + \frac{1}{r^2}\frac{\partial^2 c}{\partial \theta^2}\right) - \left(v_x \frac{\partial c}{\partial x} + v_r \frac{\partial c}{\partial r} + \frac{v_\theta}{r}\frac{\partial c}{\partial \theta}\right) \quad (7.5)$$

This equation is totally intractable but, fortunately, it may be greatly simplified for many experimental situations:

1. The system is totally symmetrical about the centre of the disc and hence the concentration is not a function of θ – all terms in $\partial c/\partial \theta$ and $\partial^2 c/\partial \theta^2$ are zero.
2. Provided the radius of the disc is small compared with that of the sheath, v_x is constant over all the disc and there is a uniform supply of reactant.
3. Since the term $(1/r)(\partial c/\partial r)$ cannot be infinite, $\partial c/\partial r$ must be zero at the centre of the disc ($r = 0$); then $\partial^2 c/\partial r^2$ must also be zero.
4. In most experiments (certainly all those covered in this section) only the steady state current density is discussed. Therefore, $\partial c/\partial t = 0$.

Making these simplifications and noting Equation (7.3), leads to the relatively straightforward expression:

$$\frac{d^2 c}{dx^2} = -\frac{0.51\omega^{3/2}}{D\nu^{1/2}}x^2\frac{dc}{dx} \quad (7.6)$$

and one approach to developing the theory for experiments at the RDE is to solve Equation (7.6) with appropriate boundary conditions at $x = 0$ and $x = \infty$.

The alternative and often more convenient approach is based upon making the assumption that only the dimension perpendicular to the disc surface is important (see previous paragraph) and then balancing the mass transport flux across the Nernst diffusion layer with the flux of electrons on the two sides of the electrode surface. Figure 7.4 is intended

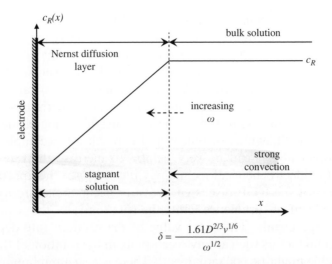

Figure 7.4 Nernst diffusion layer model for the oxidation of R to O at a rotating disc electrode. The bulk solution contains no O.

to remind the reader of the concept of a Nernst diffusion layer. The real situation where convection becomes less important as the surface is approached ($v_x \propto x^2$) is replaced by an equivalence model that invokes two discrete regions: (a) a layer immediately adjacent to the surface that is stagnant and hence only diffusion occurs and (b) outside this layer convection is strong and uniform so that no changes in concentration occur. The derivations require the dependence on experimental parameters of the thickness of the Nernst diffusion layer leading to a flux of reactant to the surface equal to that in the real situation. This is another result from Levich's quantitative description of the hydrodynamics:

$$\delta = \frac{1.61 D^{1/3} \nu^{1/6}}{\omega^{1/2}} \qquad (7.7)$$

In consequence, the mass transport coefficient is given by:

$$k_m = \frac{D^{2/3} \omega^{1/2}}{1.61 \nu^{1/6}} \qquad (7.8)$$

The Nernst diffusion layer approach works because the current density depends only on the flux of reactant to the surface and details of the concentration profile away from $x = 0$ do not need to be known. The importance of convection will increase with the rotation rate of the disc; it can be seen from Equation (7.7) that, in the Nernst diffusion layer model, this is equivalent to the boundary layer becoming thinner with rotation rate.

The above discussion, as well as that later in this chapter, assumes that we have a laminar flow regime. In a laminar flow regime, the solution moves forward in a highly organized manner so that it may be considered to consist of a sequence of separate, non-mixing elements. This is true only below a critical rotation rate. With higher rotation rates, the flow regime becomes less organized; local eddies are superimposed on the overall flow, giving rise to mixing between neighbouring elements. This type of flow is known as turbulent and is much more complex to deal with theoretically. It can be very helpful to increase the rate of mass transport in electrochemical technology but it causes the approaches to the theory for the RDE discussed above to become inappropriate. The commencement of turbulence limits the rotation rates that can be used in RDE experiments. The exact value of the rotation rate depends on experimental factors such as any eccentricity in the rotation of the disc. A typical value might be 600 radians s^{-1}. There is also a minimum rotation rate where the theory is applicable; the forced convection resulting from the rotation must be dominant compared to natural convection and non-steady state diffusion (if the voltammogram is obtained using a potential scan technique). Here a typical value might be 20 radians s^{-1}.

In this book, all the equations for the RDE are written in terms of ω, the rotation rate in radians s^{-1}. In the laboratory, the equipment usually employs rotations per minute (rpm) or rotations per second (s^{-1}), often given the symbol f. The Levich equation and related expressions would have a different numerical constant if the units of rotation rate were rotations per second or rotations per minute. Hence, to use the equations in this book, the rotation rates must be converted into radian s^{-1} using the relationships, $\omega = 2\pi f = 2\pi(\text{rpm}/60)$.

7.2.3.1 Mass Transport Control. The Nernst diffusion layer concept allows a trivial derivation of the current density at a RDE for potentials where the electrode reaction is mass transport controlled. For such potentials, the surface concentration of the electroactive species is zero so that the limiting current density is given by:

$$j_L = nFD\left(\frac{dc}{dx}\right)_{x=0} = nFD\frac{c}{\delta} \qquad (7.9)$$

and use of Equation (7.7) leads to the well known Levich equation:

$$j_L = 0.62\frac{nFD^{2/3}c}{\nu^{1/6}}\omega^{1/2} \qquad (7.10)$$

Hence the tests for mass transfer control in a RDE experiment are that the current density is independent of potential and a plot of j_L *vs*

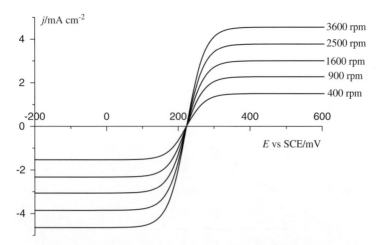

Figure 7.5 Voltammograms for a solution of 5 mM ferrocyanide + 5 mM ferricyanide in 0.5 M KCl. Au disc electrode, 298 K. The rotation rates are shown on the figure in the experimental units of rpm ($= 30\omega/\pi$).

$\omega^{1/2}$ is linear and passes through the origin (or $j_L/\omega^{1/2}$ is a constant). Furthermore, it can be seen that $j_L \propto c$ and that the diffusion coefficient can be estimated from either a j_L *vs* $\omega^{1/2}$ plot or a j_L *vs* c plot, provided the kinematic viscosity is known; in fact, an accurate value is unnecessary since, because of the power of $1/6$, an error in ν has little effect on the value of D.

Figure 7.5 shows some experimental voltammograms for a Au RDE in a solution containing equal concentrations of ferrocyanide and ferricyanide. Figure 7.6 shows the corresponding plots of the Levich equation for the limiting currents for both oxidation and reduction. The plots are linear and pass through the origin, confirming that both reactions become mass transport controlled at high overpotentials. In fact, in Figure 7.6, a single line is drawn through the data points for the oxidation and the reduction although the limiting current densities for the oxidation are slightly less than those for reduction. The diffusion coefficients for ferrocyanide and ferricyanide calculated from the data points are not quite equal, being 6.6×10^{-6} and $6.8 \times 10^{-6} \, \mathrm{cm^2 \, s^{-1}}$, respectively. Notably, in the voltammograms of Figure 7.5, close to the equilibrium potential (where $j = 0$), the current density is independent of rotation rate, confirming kinetic control. Moreover, as the over-potentials are increased into the region of mixed control, the current density becomes progressively more dependent on the rotation rate. These changes illustrate well how the RDE may be used to distinguish kinetic, mixed and mass transfer controlled regions.

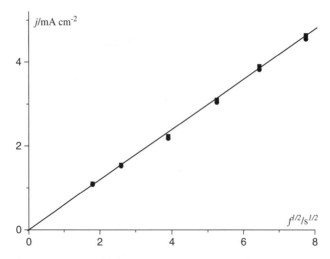

Figure 7.6   Plots of j_L *vs* $f^{1/2}$ ($f = \omega/2\pi$) from the curves of Figure 7.5 for the oxidation of ferrocyanide (●) and the reduction of ferricyanide (■); each 5 mM in 0.5 M KCl. Au disc electrode, 298 K. A single line is drawn through the data but the diffusion coefficients were estimated from the two sets of data points.

7.2.3.2   Reversible Electron Transfer.   In Section 1.4.2, the general form of a voltammogram at a rotating disc was deduced by consideration of the concentration profiles within the Nernst diffusion layer model. It was concluded that the response was sigmoidal. At any rotation rate, the Nernst diffusion layer thickness is a constant and hence the current density depends only on the surface concentration of the reactant and hence on the applied potential. For the oxidation of R to O, making the potential more positive will cause the surface concentration of R to decrease and the current density to increase. This change continues until the surface concentration reaches zero, and since it cannot change further, the current density reaches a plateau value. The exact shape of the voltammogram therefore depends on how the surface concentration of R changes with potential.

If the electron-transfer reaction:

$$O + ne^- \rightleftharpoons R \tag{7.11}$$

is rapid (*i.e.* the standard rate constant is large), the concentrations of O and R at the surface will have their equilibrium values at all potentials and rotation rates and the voltammogram will have the shape appropriate to a reversible reaction. Well positive and negative to the equilibrium potential, the reaction will be mass transport controlled and the limiting current densities will be proportional to the square root of the rotation rate, see Equation (7.10). Even at potentials in the rising portion of the voltammogram, a plot of j_L vs $\omega^{1/2}$ will be linear but the slope of the plot will be lower than in the limiting current region. Since the ratio of the surface concentrations of O and R is always in equilibrium, the complete shape of the voltammogram can be found from the Nernst equation:

$$E = E_e^\circ + \frac{2.3RT}{nF} \log\left(\frac{c_O}{c_R}\right)_{x=0} \qquad (7.12)$$

written in a form to emphasize that the electrode potential determines the concentrations at the surface. The ratio of the concentrations at the surface may be found by manipulation of the following equations resulting from the Nernst diffusion layer model. At any potential, the current density can be written in terms of a flux of R or O across the Nernst diffusion layer, *i.e.*:

$$j = nFD\frac{c_R - (c_R)_{x=0}}{\delta} \qquad (7.13)$$

and:

$$j = -nFD\frac{c_O - (c_O)_{x=0}}{\delta} \qquad (7.14)$$

In the anodic and cathodic limiting current regions the surface concentration of R and O, respectively, are zero and hence:

$$j_L^A = \frac{nFDc_R}{\delta} \qquad (7.15)$$

and:

$$j_L^C = -\frac{nFDc_O}{\delta} \qquad (7.16)$$

Algebraic manipulation of these expressions and substitution for the surface concentrations in Equation (7.12) leads to:

$$E = E_e^\circ + \frac{2.3RT}{nF} \log\frac{j - j_L^C}{j_L^A - j} \qquad (7.17)$$

Equation (7.17) provides a test for a reversible electrode reaction; a plot of E vs $\log[(j - j_L^C)/(j_L^A - j)]$ should be linear with a slope of $2.3RT/nF$ and the intercept gives a value for the formal potential of the couple. It also allows the voltammograms at all rotation rates to be collapsed onto a single presentation. Of course, the above derivation applies to a solution that initially contains both O and R. If the solution under study contains only O or R, then the appropriate expression can be written by recognizing that one limiting current will be zero.

7.2.3.3 Irreversible Electron Transfer. When the kinetics of electron transfer at the electrode are insufficiently fast to maintain the concentrations of O and R at the surface in equilibrium it is necessary to consider the kinetics of electron transfer. An overpotential will be necessary to drive the reactions and, depending on the kinetics, the voltammogram will either be less steep or, if the standard rate constant is low, change from a single wave into separate oxidation and reduction waves spaced by a potential range where the current density is very low.

The RDE provides a procedure to determine the kinetic parameters when the electrode reaction is irreversible. In addition, for a range of standard rate constants, there is the possibility of the voltammogram changing from that for a reversible reaction to that for an irreversible reaction by increasing the rotation rate of the disc.

If the current density is fully kinetically controlled, the current density will be independent of the rotation rate of the disc; the rate constant for electron transfer at each potential can be calculated directly from the current density. The determination of the kinetics can, however, be extended into the region of mixed control using the RDE. The conceptual approach will be (a) at each potential to measure the current density as a function of the rotation rate of the disc and (b) to remove any possible influence of mass transport by deducing the current density at infinite rotation rate where k_m is infinite. The resulting rate constant at each potential can then be plotted as a function of potential to obtain the standard rate constant and transfer coefficient (see below). To employ this approach, it is advantageous to employ a linear plot to extrapolate to infinite rotation rate.

The appropriate equation can again be deduced from the Nernst diffusion layer model. Firstly, at any potential, for mass balance the fluxes of O and R at the surface must be equal, *i.e.*:

$$-k_m\left[c_O - (c_O)_{x=0}\right] = k_m\left[c_R - (c_R)_{x=0}\right] \tag{7.18}$$

Secondly, the current density at any potential can be written in terms of either the mass flux through the Nernst diffusion layer or the kinetics of electron transfer, *i.e.*:

$$j = nFk_m\left[c_R - (c_R)_{x=0}\right] \tag{7.19}$$

and:

$$j = nF\left[k_a(c_R)_{x=0} - k_c(c_O)_{x=0}\right] \tag{7.20}$$

Combining the equations so as to eliminate the two surface concentrations leads to:

$$\frac{1}{j} = \frac{1}{nF(k_a c_R - k_c c_O)} + \frac{k_a + k_c}{nF k_m(k_a c_R - k_c c_O)} \tag{7.21}$$

Finally, to obtain the desired expression, it is necessary to note that $k_m = D/\delta$ and that the thickness of the Nernst diffusion layer is given by Equation (7.7):

$$\frac{1}{j} = \frac{1}{nF(k_a c_R - k_c c_O)} + \frac{1.61 \nu^{1/6}(k_a + k_c)}{nF D^{2/3}(k_a c_R - k_c c_O)} \frac{1}{\omega^{1/2}} \tag{7.22}$$

In practice, it is necessary only to consider both partial oxidation and partial reduction current densities over a narrow potential range close to the equilibrium potential (*ca.* within 60 mV of E_e). At all other potentials, one partial current density will dominate totally and Equation (7.22) can be simplified. At positive overpotentials, $k_a c_R \gg k_c c_O$ and:

$$\frac{1}{j} = \frac{1}{nF k_a c_R} + \frac{1.61 \nu^{1/6}}{nF D^{2/3} c_R} \frac{1}{\omega^{1/2}} \tag{7.23}$$

and a similar simplification is possible if the initial solution contains only R. At negative overpotentials or when the solution contains only O:

$$-\frac{1}{j} = \frac{1}{nF k_c c_O} + \frac{1.61 \nu^{1/6}}{nF D^{2/3} c_O} \frac{1}{\omega^{1/2}} \tag{7.24}$$

Each of Equations (7.22–7.24) indicate the way to obtain the current density under conditions of an infinite rate of convection. The current density should be determined as a function of the rotation rate, ω, and the data replotted as $1/j$ *vs* $1/\omega^{1/2}$. The intercept at $1/\omega^{1/2} = 0$ is the inverse of the current density at infinite rotation rate and reflects only the kinetics of electron transfer. The rate constant for electron transfer can then be plotted as a function of potential. In the potential region where both the forward and back electron-transfer reaction should be considered, k_a and k_c can be obtained using both the slope and intercept of the $1/j$ *vs* $1/\omega^{1/2}$ plot, see Equation (7.22).

This procedure can be illustrated by further analysis of the voltammograms in Figure 7.5. Figure 7.7 shows typical $1/j$ *vs* $1/f^{1/2}$ plots

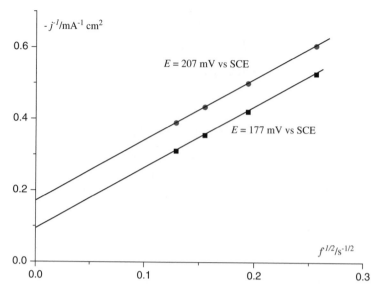

Figure 7.7 Typical plots of Equation (7.24), $1/j$ vs $1/f^{1/2}$ ($f = \omega/2\pi$) for determination of the rate constant for the reduction of ferricyanide to ferrocyanide as a function of potential; k_c is determined from the intercepts using Equation (7.25). The solution contained 5 mM ferrocyanide + 5 mM ferricyanide + 0.5 M KCl at 298 K.

for two potentials in the potential range between the plateaux regions (remember that f is the frequency of rotation in s^{-1} and $f = \omega/2\pi$). The intercepts are then used to calculate the rate constants for electron transfer at each potential using:

$$\left(\frac{1}{j}\right)_{\frac{1}{\omega^{1/2}}=0} = \frac{1}{nFk_a c_R} \tag{7.25}$$

(written for oxidation). The rate constants are then plotted as log k_a and log k_c vs E (Figure 7.8). The intersection of the two lines gives values for log k_s and the formal potential for the couple being studied.

An alternative approach to determining the kinetics of electron transfer is based on the analysis of the shape of the voltammetric wave as a function of the rotation rate of the disc. Figure 7.9 illustrates the way in which the shape of the voltammetric wave for an electrode reaction varies with the standard rate constant for electron transfer (the rotation rate of the disc is kept constant). With a decrease in the standard rate constant, the overpotential required to drive the reaction at each current density increases and the wave shifts and becomes less

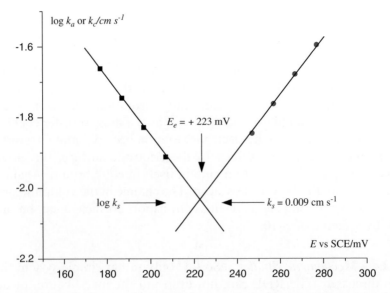

Figure 7.8 Analysis of experimental data in Figures 7.5 and 7.7 to obtain the formal potential and standard rate constant for the ferricyanide/ferrocyanide couple. The solution contains 5 mM ferricyanide + 5 mM ferrocyanide in 1 M KCl at 298 K ($\bullet$ = data for log k_a; $\blacksquare$ = data for log k_c).

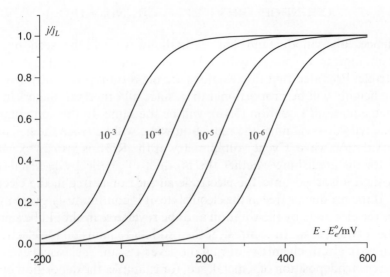

Figure 7.9 Variation of the shape of a voltammogram for the oxidation of R to O with the standard rate constant for electron transfer (the values in cm s^{-1} are shown on the figure). The curves are drawn for a rotation rate of 900 rpm ($k_m = 9.5 \times 10^{-4}$ cm s^{-1}) and for a solution containing R but no O.

steep. Of course, the limiting current density is independent of the standard rate constant for electron transfer. Hence, the standard rate constant is readily found by fitting the experimental response to a set of theoretical curves such as those shown in the figure. A similar figure could be constructed for the variation of the voltammogram with increase in the mass transfer coefficient (*i.e.* rotation rate) for a reaction with a particular standard rate constant; as the mass transfer coefficient is increased ($k_m \propto \omega^{1/2}$) the electrode reaction becomes more irreversible (*i.e.* as more reactant is forced to the electrode surface, the electron transfer couple is less able to maintain itself in equilibrium). Again, the wave will shift and become less steep. The change in the voltammograms will, however, be smaller as the mass transport coefficient can be varied only by a factor of ~ 10.

7.2.3.4 Mechanistic Studies.   So far, only rather simple systems have been discussed. The RDE can, however, aid the investigation of more complex systems. Firstly, it should be recognized that the voltammetry for some reactants will show more than one wave. For example, in aqueous acid, phenylhydroxylamine will both oxidize and reduce:

$$C_6H_5NHOH - 2e^- \longrightarrow C_6H_5N = O + 2H^+ \qquad (7.26)$$

$$C_6H_5NHOH + 4H^+ + 4e^- \longrightarrow C_6H_5NH_2 + H_2O \qquad (7.27)$$

Hence, one anodic and one cathodic wave will be seen on the voltammogram of phenylhydroxylamine at an amalgamated Cu disc electrode. Provided that the reactions are mass transport controlled, the wave heights will be proportional to n, since all other parameters in the Levich equation, Equation (7.10), will be the same. In the voltammetry of phenylhydroxylamine, the reduction wave will be twice the height of the oxidation wave. Cyclic voltammetry at the RDE is also an excellent tool for distinguishing whether the product of an electrode reaction is deposited/adsorbed onto the electrode surface or is free in the electrolyte. If the product is free in the electrolyte, it is continuously swept away from the electrode by the rotation and the reverse scan should be similar to the forward scan. In contrast, if the product remains on the electrode, it can have electrochemistry during the reverse scan. A classical example is the electrodeposition of a metal, see, for example, the deposition of Au onto a vitreous carbon disc (Figure 7.10). The forward scan shows a cathodic wave for the reduction of Au(III) to Au metal and variation of the rotation rate would show that this reaction is mass transfer controlled ($j_L \propto \omega^{1/2}$). On the reverse the metal deposition continues

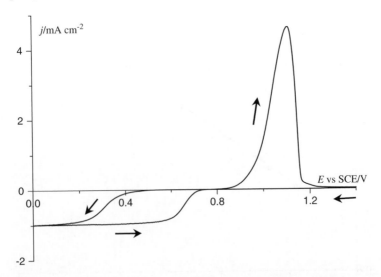

Figure 7.10 Cyclic voltammogram recorded for the deposition and dissolution of gold onto a vitreous carbon disc electrode (diameter 0.3 cm) from a solution containing 1 mM HAuCl$_4$ + 0.1 M HCl, deoxygenated with Ar. Rotation rate: 900 rpm, potential scan rate 1 mV s^{-1}, 298 K.

until a critical potential, in fact well positive to where deposition occurs on the forward scan; this difference arises because of the need for nucleation of the gold phase on the vitreous carbon surface on the forward scan. At more positive potentials an anodic stripping peak is observed. The oxidation process is not limited by mass transport as the gold has been plated onto the electrode surface and hence large current densities are obtained. On the other hand, the oxidation charge is limited by the amount of gold electroplated. These two factors together, lead to the characteristic large, symmetrical anodic peak. In addition, even on the reverse scan, the reduction wave and oxidation peak are separated by a potential range where the current density is very low. This indicates that the kinetics of the Au(III)/Au couple are slow and therefore that overpotentials are necessary to drive both reduction and oxidation.

The RDE can also be used to study homogeneous chemistry. For example, when carbon dioxide is dissolved in a neutral aqueous electrolyte the solution becomes slightly acidic (pH ∼4) and a low concentration (<0.2 mM) of carbonic acid is formed but the dominant species is CO$_2$ (∼28 mM at 298 K and ∼12 mM at 333 K). A voltammogram at a Pt RDE shows a well-defined reduction wave for the hydrogen evolution reaction; however, analysis of the limiting current as a function of rotation rate (Figure 7.11) shows that the j_L *vs* $\omega^{1/2}$ plot

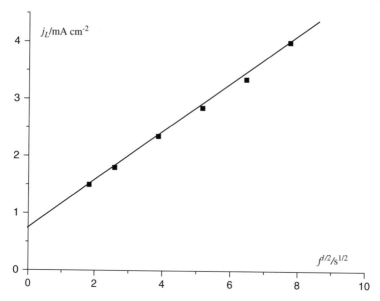

Figure 7.11   Plot of limiting current density *versus* the square root of the frequency of rotation for data taken from voltammograms for a Pt RDE in an aqueous solution containing 0.5 M NaCl saturated with CO_2; temperature 333 K. Data taken from M. J. Medeiros, D. Pletcher and D. Sidorin, *J. Electroanal. Chem.*, 2008, **619/620**, 83.

is linear (confirming that mass transport controlled reactions are occurring) but there is also an intercept on the limiting current axis (*i.e.* there is also a reaction independent of the rate of mass transport). This observation can be explained by the recognition that the two reactions:

$$2H^+ + 2e^- \longrightarrow H_2 \tag{7.28}$$

$$2H_2CO_3 + 2e^- \longrightarrow H_2 + 2HCO_3^- \tag{7.29}$$

are mass transport controlled but that current density is amplified by the slow continuous hydration of carbon dioxide:

$$CO_2 + H_2O \longrightarrow H_2CO_3 \tag{7.30}$$

The carbonic acid formed in this hydration reaction close to the surface can reduce, Reaction (7.29), and this leads to a continuous interconversion within the reaction layer. Since there is a high concentration of carbon dioxide within the reaction layer and the intercept is much less than the current density predicted for the mass transport controlled flux of CO_2, the intercept is a measure of the rate of the hydration reaction.

In fact, the rate constant for Reaction (7.30) can be shown to be 0.04 and $0.9 \, s^{-1}$ at 298 and 333 K, respectively.

In complex reactions such as the reduction of oxygen (Section 5.4) the number of electrons transferred can be a function of the mass transfer coefficient. Then, with a RDE experiment, the plot of j_L *vs* $\omega^{1/2}$ will be nonlinear; n is proportional to $j_L/\omega^{1/2}$ and hence a suitable data analysis is to plot $j_L/\omega^{1/2}$ as a function of k_m. In the case of oxygen reduction, $j_L/\omega^{1/2}$ will decrease with increasing rotation rate if hydrogen peroxide is formed as an intermediate, since increasing the rotation rate will lead to more of the intermediate being swept away from the surface before further reduction/disproportionation can occur.

7.2.4 Rotating Ring Disc Electrodes (RRDE)

Only a limited range of homogeneous chemistry can be investigated using a RDE. Usually, the chemistry of species formed in an electrode reaction must be studied in an experiment with two parts. In the first, the species is formed in a controlled manner, while in the second, the electro-activity of the species formed is effectively used to examine the fraction of the species not undergoing reaction. To allow such experiments, the rotating ring disc electrode (RRDE) was developed. In this structure the disc electrode is surrounded by a ring electrode with a thin insulating gap between them (Figure 7.12). The two electrodes are electrically isolated and may be controlled separately. The flow pattern produced by rotation of the RRDE is identical to that of the RDE; the solution is drawn up to the disc and then thrown out across the surface of the rotating structure. Therefore, the ring electrode is downstream to the disc electrode. Some but not all the species formed at the disc will pass

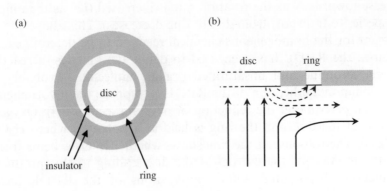

(a)   (b)

disc   ring

disc

insulator

ring

Figure 7.12 (*a*) View of the rotating ring disc electrode (RRDE) from below and (*b*) solution flow pattern at the RRDE.

to the ring (Figure 7.12b). The fraction of the species reaching the ring will depend on the characteristics of the RRDE (*i.e.* the radius of the disc and the inner and outer radii of the ring) as well as the stability in the electrolyte of the product from the disc and the rotation rate, as this determines the time for the species to transport from the disc to the ring.

It must be recognized that not all the species, even a completely stable species, formed at the disc will reach the ring. Because of the flow pattern at the rotating surface, some of the species will bypass the ring and pass into the bulk solution. The collection efficiency, N, is defined as the fraction of a completely stable species formed at the disc that is detected at the ring. For a system:

Disc $O + ne^- \longrightarrow R$

Ring $R - ne^- \longrightarrow O$

where O and R are completely stable (*e.g.* ferricyanide and ferrocyanide) and the potentials of the disc and ring are held at values where the reactions are mass transport controlled:

$$N = -\frac{I_{\text{ring}}}{I_{\text{disc}}} \tag{7.31}$$

Although both currents increase with rotation rate, the ratio is independent of the rotation rate. While theoretical expressions relating N to the disc and ring dimensions are in the literature, N is usually found by experiment (typical values are 0.2–0.4).

If the product of the disc reaction, R, participates in a homogeneous chemical reaction, $-I_{\text{ring}}/I_{\text{disc}}$ will fall below N and then it will tend to increase towards N as the rotation rate is increased (*i.e.* as the time for the species to transport from disc to ring decreases). This allows the rate constant for the homogeneous chemical reaction to be determined. For example, the RRDE has been used to define the kinetics of reactions between bromine and unsaturated organic molecules. Probably, the most common application of the RRDE is to probe the involvement of hydrogen peroxide in the reduction of oxygen (Section 5.4). Oxygen is reduced at the disc and the ring is held at a potential where H_2O_2 is oxidized. The response at the ring shows whether H_2O_2 is being formed and the fraction of the current at the disc leading to this product as a function of potential. Other applications of the RRDE include distinguishing between surface-bound and solution free products from reactions at the disc; this experiment is particularly helpful in corrosion where it allows the partition of the corrosion current between corrosion film formation and solution free ions.

Since the disc and the ring are independently controlled, several types of experiment are possible:

1. The disc is held at a constant potential where the reaction of interest occurs while a voltammogram is recorded at the ring. The potentials for waves in the voltammogram allow intermediates/ products free in the solution phase to be identified while the magnitudes of the limiting currents give information about the yields of these species and/or their stability. The variation of the voltammogram with rotation rate allows the study of the kinetics of the homogeneous reaction.
2. A voltammogram is recorded at the disc while the potential of the ring is held constant. This allows the recognition of a particular intermediate and definition of the potential range where it is formed.
3. Both ring and disc electrodes are held at constant potentials. This can be the best mode of operation for determining N and obtaining quantitative kinetics.

The RRDE allows many elegant experiments that give conclusive results. The drawback is that RRDEs with a sufficiently thin disc/ring gap are difficult to fabricate (and hence expensive to purchase) and they must also be maintained in an undamaged state (polished surfaces and no damage to the insulating gap).

7.3 NON-STEADY STATE TECHNIQUES

Almost always, non-steady state experiments are carried out under conditions of linear diffusion to a plane electrode. The experiments are carried out in unstirred solution, preferably in a thermostat and on a timescale where natural convection is negligible. In addition, an excess of an inert electrolyte is added so that migration does not contribute significantly to the mass transport of the electroactive species. Hence, diffusion is the only significant mode of mass transport. Because of the timescale, the experiments take place while the diffusion layer is developing, *i.e.* the concentrations of the reactant, intermediates and products are a function of both time and distance from the electrode surface. The selection of the conditions so that the experiments may be described by linear diffusion to a plane electrode allows the use of Fick's laws in one dimension to quantify the mass transport regime.

7.3.1 Potential Step Experiments

Although potential step experiments can be a good way to obtain quantitative data when a system is understood, qualitative information also usually requires some analysis of the shape of the current *vs* time transient. Hence, it is a less used technique than, for example, cyclic voltammetry that allows qualitative interpretation without mathematical analysis. Moreover, the experimental current will have a contribution from double layer charging (Section 6.2.2) that is additive to the Faradaic current and this distorts the response (particularly at short times) without obviously changing the shape of falling transients. Experimentally, the timescale of potential step experiments is limited by charging currents at short times and interference from natural convection at long times; typically, the useful range is 1 ms to 10 s. In this book, the experiments are, perhaps, most important for their contribution to understanding non-steady state diffusion.

7.3.1.1 Diffusion Control. In Section 1.4.1, one potential step experiment has been discussed. The solution contained O and R as well as an excess of electrolyte and the potential of an inert electrode was stepped from a value where no current flows (here, the equilibrium potential for the couple O/R) to a high overpotential where the oxidation of R to O is diffusion controlled. It is, however, appropriate to reiterate the main arguments here.

The change in potential will lead to an instantaneous change in the concentration of R at the surface from the bulk value, c, to zero and an anodic current will immediately be observed. The change in concentration caused by electron transfer is possible only on the electrode surface, $x = 0$. This change, however, inevitably leads to differences in concentration within the solution very close to the electrode surface. As a result, diffusion of R towards the surface occurs. This, in turn, produces concentration differences away from the surface and hence to the expansion of the diffusion layer into solution. Figure 7.13 illustrates the development of the concentration profiles with time. With increasing time (a) the diffusion layer becomes thicker, (b) the gradients in concentration decrease with time at all values of x (this is the role of diffusion) and (c) the flux of R to the surface (and hence the current density) decreases with time – a falling current density *versus* time is observed.

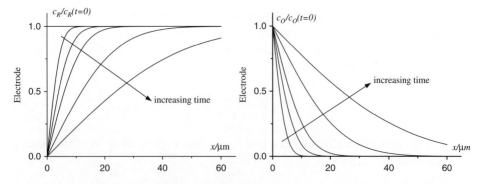

Figure 7.13 Concentration profiles resulting from a potential step experiment where the potential is stepped from a value where the current density is zero to one where the oxidation of R to O is diffusion controlled. The initial solution contains R but not O. The profiles are shown for times in the range 10 ms–1 s for species where $D = 10^{-5}$ cm^2 s^{-1}.

The exact form of the transient is found by using Laplace transform techniques to solve Fick's second law for the reactant R:

$$\frac{\partial c_R}{\partial t} = D\frac{\partial^2 c_R}{\partial x^2} \tag{7.32}$$

with the appropriate initial and boundary conditions, *i.e.*:

$$\text{at } t = 0 \text{ and at all } x, \quad c_R(x,0) = c_R \tag{7.33}$$

and for $t > 0$, at $x = 0$, $c_R(0,t) = 0$ and at $x = \infty$, $c_R(\infty,t) = c_R$ (7.34)

and noting that the flux of electrons into the electrode and of R to the surface must balance:

$$\frac{j}{nF} = D\left(\frac{\partial c}{\partial x}\right)_{x=0} \tag{7.35}$$

leads to the Cottrell equation:

$$j = \frac{nFD^{1/2}c}{\pi^{1/2}t^{1/2}} \tag{7.36}$$

The falling current density *versus* time transient will follow this equation. Essentially, the Cottrell equation is a test for diffusion control. The shape of the transient may be tested by (a) plotting j vs $t^{1/2}$ and checking whether a straight line passing through the origin is obtained, (b) demonstrating that $jt^{1/2}$ is a constant or (c) plotting Equation (7.36) for various values of D and fitting to the experimental data. Each will lead to a value for the diffusion coefficient.

Some workers prefer to analyse charge *vs* time rather than current *vs* time. Integration of the current as a function of time is readily carried out electronically and the theoretical equation is easily obtained by integration of Equation (7.36):

$$q = \frac{2nFD^{1/2}c}{\pi^{1/2}}t^{1/2} \qquad (7.37)$$

The test for diffusion control is a linear plot of q *vs* $t^{1/2}$ passing through the origin.

7.3.1.2 Electron-transfer Control. Stepping the potential into the range for total electron-transfer control corresponds to changing the surface concentration of the reactant from the bulk value by a very small amount, *e.g.* from c_R to $0.999c_R$ (remember that the current density may be less than the diffusion controlled current density by a factor of $> 10^3$ and that a Tafel plot should be over several orders of magnitude of current density). On a plot of the concentration profile it will appear that $(c)_{x=0} = c$ and, as a result, the observed current density should be a constant, low value. This is, however, also the situation where the charging current will distort the transient most strongly and, in practice, a falling transient will therefore be seen at short times.

7.3.1.3 Mixed Control. Stepping into the potential range for mixed control corresponds to changing the surface concentration of the reactant from the bulk value by a significant amount but not to zero (since this would correspond to diffusion control). A discussion of the development of the concentration profiles largely parallels that of the case of diffusion control (Figure 7.14). Immediately at the step in potential, a substantial change occurs to the surface concentration of R (but only at the surface). This change, however, leads to concentration differences in solution and diffusion occurs to minimize the differences in concentration and to push the diffusion layer out into the solution. The conclusion must be that a falling transient will again be observed but, at all times, the flux of R to the surface, and hence the current density, will be lower than those for diffusion control. It can further be seen in Figure 7.14 that, following the immediate change in surface concentration on stepping the potential, the surface concentration continues to change for a time. This is because the electron-transfer reaction is not fast and the equilibrium surface concentrations of R and O take time to be established.

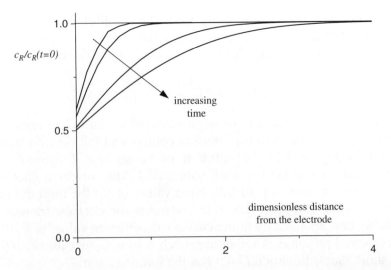

Figure 7.14 Concentration profiles for an experiment where the potential is stepped from a value where the current is zero to one where the oxidation of R to O is under mixed control. The initial solution contains R but not O.

The exact form of the transient results from the solution of a pair of partial differential equations (because the boundary conditions involve the concentrations of both O and R):

$$\frac{\partial c_R(x,t)}{\partial t} = D\frac{\partial^2 c_R(x,t)}{\partial x^2} \quad \text{and} \quad \frac{\partial c_O(x,t)}{\partial t} = D\frac{\partial^2 c_O(x,t)}{\partial x^2} \tag{7.38}$$

with the initial conditions, at $t=0$:

$$c_R(x,t) = c_R \quad \text{and} \quad c_O(x,t) = c_O \tag{7.39}$$

and the boundary conditions, at $x=\infty$: $c_R(x,t) = c_R$ and $c_O(x,t) = c_O$ and at $x=0$:

$$D\left(\frac{\partial c_R}{\partial x}\right)_{x=0} = -D\left(\frac{\partial c_O}{\partial x}\right)_{x=0} = k_a c_R(0,t) - k_c c_O(0,t) \tag{7.40}$$

Since the surface concentrations are unknown, the boundary conditions at the surface arise by equating the two diffusion fluxes and the kinetic flux at the surface. These more complex boundary conditions lead to more complex mathematics, but again Laplace transform methods may be used to obtain the form of the transient:

$$j = nFk_a c_R \exp\frac{(k_a + k_c)^2 t}{D} \text{ erfc}\frac{(k_a + k_c)t^{1/2}}{D^{1/2}} \tag{7.41}$$

At a potential sufficiently away from the equilibrium potential (*i.e.* $\eta > 50\,\text{mV}$) that one of the potential dependent rate constants is far larger than the other, simplification is possible; at positive overpotentials:

$$j = nFk_a c_R \exp\frac{k_a^2 t}{D}\ \text{erfc}\ \frac{k_a t^{1/2}}{D^{1/2}} \tag{7.42}$$

with an equivalent equation for negative overpotentials. This response is compared to the transient for diffusion control and full electron transfer control in Figure 7.15. Substitution of values into Equation (7.42) would confirm that (a) for small values of k_a the current is effectively independent of time and (b) with large values of k_a, the response tends to the Cottrell response. The rate constants for electron transfer are probably best obtained by fitting curves with different values of k_a to the experimental response. Another approach is to note that for $(k_a^2 t/D) \ll 1$ (*i.e.* short times) Equation (7.42) has the limiting form:

$$j = nFk_a c_R\left[1 - \frac{2k_a}{\pi^{1/2}D^{1/2}}t^{1/2}\right] \tag{7.43}$$

and the rate constant can be found from the intercept or the slope of a plot of j *vs* $t^{1/2}$. The intercept at $t=0$ corresponds to an extrapolation to the hypothetical situation where the rate of diffusion is infinite (in

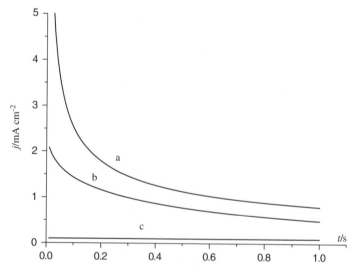

Figure 7.15 Current density *versus* time response to a potential step experiment (a) at a potential where the electrode reaction is diffusion controlled – Equation (7.36), (b) at a potential where there is mixed control, Equation (7.42), and (c) at a potential where the oxidation is fully electron-transfer controlled; $c = 5\,\text{mM}$, $D = 10^{-5}\,\text{cm}^2\,\text{s}^{-1}$, $k_a = 5 \times 10^{-3}\,\text{cm}\,\text{s}^{-1}$.

Figure 7.14, if one goes backwards in time to $t = 0$, the profile will have infinite slope). In practice, Equation (7.43) is difficult to use because it necessitates the measurement of the Faradaic current density on a timescale where interference from the charging current is most marked. Indeed, it can be seen that the higher the value of k_a the shorter must be the experimental times for Equation (7.43) to be valid.

Equation (7.42) has another limiting form. When $k_a^2 t/D) \gg 1$, it collapses to the Cottrell equation, (7.37). In summary, the rate of diffusion is a function of time, decreasing as $1/t^{1/2}$. Hence, electron transfer control is most likely at short times and diffusion control at long times, but the faster the kinetics of electron transfer, the shorter is the timescale essential for observing some kinetic control. The potential step approach will be used to determine the rate constant for electron transfer in two distinct situations: (a) to extend a Tafel plot from the potential range where steady state data can be used to determine kinetic information into the potential range of mixed control and (b) when the standard rate constant is sufficiently high that, in the steady state, the reaction appears reversible.

Again, the measurement of charge *vs* time can be advantageous. After integration of the current transient, the information from short times is retained through a longer timescale. Therefore, the kinetic parameters influence the response over a much wider timescale. Extrapolation of the long timescale data back towards $t = 0$ will show a time range where the charge is less than predicted for diffusion control. Integration of Equation (7.42) leads to an expression of the form:

$$q = \frac{4nFk_a c_R}{\pi^{1/2}} \left(T^{1/2} t^{1/2} - T \right) \qquad (7.44)$$

Figure 7.16 shows a plot of q *vs* $t^{1/2}$. The rate constant is obtained from the intercept and slope.

Once again, it must be emphasized that k_a and k_c are rate constants at one particular potential. The fundamental kinetic parameters, the standard rate constant (or exchange current density) and transfer coefficient are obtained by determining k_a and k_c at a series of potentials and then plotting $\log k_a$ and $\log k_c$ *vs* E as in a Tafel plot.

7.3.1.4 Nucleation and Growth of New Phases. The formation of a new phase on an electrode involves nucleation and growth (Section 1.9). When the potential of the substrate is stepped to a potential where these processes are initiated, the response will initially be a current that

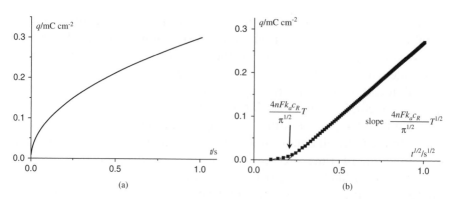

Figure 7.16 (*a*) Typical chronocoulometric response following a potential step for the oxidation of R to O, (*b*) analysis of a *j* vs *t* transient at a potential for mixed control using Equation (7.44) – the rate constant for electron transfer can be found using the slope and intercept.

increases with time; the number of nuclei may increase with time and the surface area of each centre available for further deposition will certainly increase. At longer times, the current may go through a maximum or reach a plateau value as the centres overlap into a continuous layer. The exact form of the transient depends on whether:

- nucleation is instantaneous or progressive;
- growth of nuclei is in two or three dimensions and the geometry of three-dimensional growth;
- growth occurs under electron transfer or mass transport control.

For detailed discussion of these possibilities, the reader should look at the texts at the end of the chapter. Here, we stress the importance of considering the whole response for both qualitative and quantitative interpretation in terms of detailed mechanisms.

Figure 7.17 illustrates the responses that can be obtained. In these experiments, palladium was being plated onto a vitreous carbon disc from an electrolyte containing 50 mM $PdCl_2$ and 1 M KCl at pH 3.5 and the deposition potential was varied. All the responses show an early falling component due to charging of the double layer. After this has decayed to a low value, rising current *vs* time transients are observed as nuclei of palladium are formed and grow. As expected the rate of both nucleation and growth increase with applied overpotential and hence the timescale of the rise decreases as the potential is made more negative. At lower overpotentials, the current is still rising after one second but at the more negative overpotentials the transients coalesce beyond a peak and

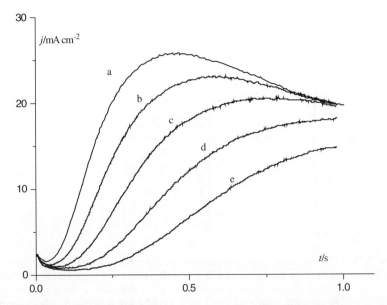

Figure 7.17 Current density *vs* time transients for the nucleation and growth of a palladium deposit on a vitreous carbon electrode. Solution: 50 mM $PdCl_2 + 1$ M KCl, pH 3.5. Potential steps from $+300$ mV to (a) -30, (b) -20, (c) -10, (d) 0 and (e) $+10$ mV *vs* SCE.

there is a smoothly falling portion to the transients. In fact, this falling component continues for many seconds and it can be fitted to the Cottrell equation, showing that, in this time range, the Pd layer is thickening under diffusion control and that the electrode models as linear diffusion to a plane electrode. The initial Pd nuclei have grown and overlapped to give a continuous metal layer.

7.3.1.5 Coupled Chemical Reactions. The study of homogeneous chemical reactions following electron transfer requires the use of a double potential step. Consider a simple ec mechanism such as:

$$R - ne^- \rightleftharpoons O \tag{7.45}$$

$$O \xrightarrow{k} P \tag{7.46}$$

where the product is unstable in the electrolyte. The appropriate double potential step experiment to determine the kinetics of the following chemical reaction is: (a) the potential is stepped from a potential where $j = 0$ to a potential where the oxidation of R to O occurs; (b) after a period, τ, the potential is then stepped to a value where the reduction of

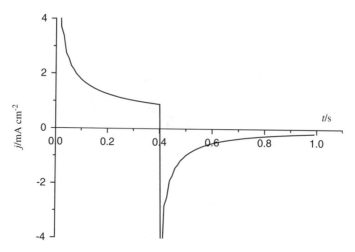

Figure 7.18 Current density *vs* time response to a double potential step for the system set out in Equation (7.45) where R and O are both stable. At the first potential, the oxidation of R to O is diffusion controlled and at the second the reduction of O to R is diffusion controlled. The response is calculated for $c_O = 2$ mM, $n = 2$ and $D = 10^{-5}$ cm^2 s^{-1}.

O to R occurs. It simplifies the mathematical description of the experiment if the potentials are selected so that the electrode reactions are diffusion controlled. Figure 7.18 shows the response for the case where $k = 0$. During the first pulse, a falling oxidation current obeying the Cottrell equation will be observed, while for $t > \tau$ a falling reduction current will result from the diffusion controlled reduction of O remaining close to the electrode. The ratio of charges, q_c/q_a, will always be much less than one since O is diffusing away from the electrode throughout both pulses; in fact when $k = 0$, $q_c/q_a = 0.23$. When $k > 0$, the chemical step (7.46) will consume O throughout the experiment and a concentration profile for O would show less O at all values of x. The second pulse effectively analyses the form of the concentration profile at $t = \tau$. Hence, the transient during the first pulse will be relatively unaffected but the current densities at all times (as well as the total charge) will be less during the second pulse. The current density during the second pulse is clearly a function of the chemical rate constant, k, and its value can therefore be determined by looking at the ratios of current densities or charges during the two pulses. The most reliable data will result when the length of the first pulse is comparable to the half-life of the intermediate, O. More complex mechanisms, *e.g.* ece reactions, can also be investigated provided it is possible to solve the partial differential equations, including the terms describing the chemistry, or digital simulation is used to analyse the data. But it should be stressed that the double potential step procedure should be used to obtain quantitative data only once the mechanism is fully understood because all mechanisms will give qualitatively the same response – falling transients during both steps. Once the mechanism is known, however, the double potential step can be a reliable way to obtain rate constants in the range 0.1–1000 s^{-1}. This range is again determined by experimental limitations, *i.e.* double layer charging at short times and natural convection at long times.

7.3.2 Cyclic Voltammetry

Cyclic voltammetry is now a very popular technique, particularly powerful for building up an understanding of new systems. Once the cell is set up, each experiment takes only a few seconds and the data are presented in a form that allows rapid, qualitative interpretation without recourse to calculations. In consequence, the insight gained from one experiment may immediately be used in the design of the next. The methods for obtaining quantitative kinetic data will also be developed although the experimenter must always be aware of the distortions arising from *IR* drop and double layer charging (Chapter 6).

Figure 7.19a shows the potential/time waveform used for cyclic voltammetry. The response is generally presented as a plot of current density (or current) *vs* potential (Figure 7.19b). In the interpretation of the response, it will always be necessary to remember that the potential axis could equally be labelled as a time axis (there is a linear relationship between E and t). In the experiment, the potential is swept through the

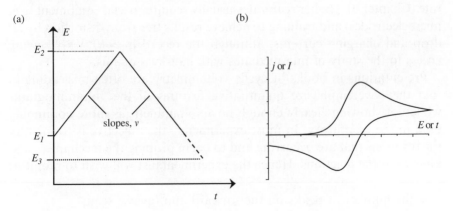

Figure 7.19 Experiment of cyclic voltammetry: (*a*) the potential–time profile employed and (*b*) the measured response.

potential range where an electrode reaction occurs, before the direction of scan is reversed to determine whether the product of electron transfer is stable and whether other electroactive species are formed. Under the control of the experimenter are:

- the potential limits – E_1, E_2 and E_3 and the direction of the initial scan;
- the potential scan rate, ν;
- whether only a first scan or multiple cycles are recorded.

and the information gained by varying these parameters may be supplemented by variation in other conditions, including the concentration of reactant, electrode material, pH and temperature. The choice of the initial potential determines the initial concentration profiles for the reactant, intermediates and products. Unless there is a specific reason, the choice should be a potential where the current density is zero; then the initial concentration profiles will show uniform concentrations throughout the diffusion layer and this concentration will depend only on the solution prepared for the experiment. The potentials will be scanned initially in a positive direction to study oxidation and negative to investigate reductions. The potential scan rate determines the timescale of the experiment, and hence the rate of non-steady state diffusion and also the timescale on which coupled chemistry is observed (in general, the voltammetry will be most revealing when the timescale of the experiment and the half-life of an intermediate are similar). Cyclic voltammetry is usually straightforward for potential scan rates in the range $25–1000 \, \text{mV s}^{-1}$; this range can be extended to $\sim 100 \, \text{V s}^{-1}$ by applying the experimental precautions necessary to overcome the dual problems of *IR* drop and charging currents, both of which increase with potential scan rate (Chapter 6). Higher scan rates usually require special equipment (*e.g.* microelectrodes) and training to achieve results free from distortion by *IR* drop and charging currents, although the record is $> 10^6 \, \text{V s}^{-1}$, giving access to the study of intermediates with half-lives of $1 \, \mu\text{s}$.

Presentations in books of cyclic voltammetry are often misleading in that they overemphasize quantitative features of the voltammograms while not showing clearly enough how voltammetry is more commonly used in the laboratory. In most experiments, the objective is to identify the reactions that are occurring and to begin probing the mechanism and kinetics of the reactions. Hence the experimentalist will wish to monitor:

- the number of peaks on the forward and reverse scan;
- the shapes of the peaks;
- peak potentials;

- peak current densities;
- charges associated with peaks and charge balance between peaks;
- differences between a first cycle, a second cycle and multiple cycles.

and particularly the way in which each of these changes with potential scan rates and potential limits. To obtain the maximum information from the experiments it is important that each response is interpreted as it is recorded and the conclusions used in the planning of the next experiment. Clearly, understanding the voltammetric response for particular types of system is essential to the interpretation in the laboratory.

7.3.2.1 Understanding Peak Shape. For each electrode reaction, a peak will be observed on the cyclic voltammogram. Two limiting types of peak shape are observed depending on whether the reaction rate is limited by non-steady state diffusion of the reactant to the electrode or by the number of surface sites in some way (*e.g.* the formation of a monolayer, the oxidation/reduction of a monolayer, the dissolution of a multilayer).

First, we consider a linear sweep experiment for the reaction $R \rightarrow O$ where both R and O are dissolved in the electrolyte. Figure 7.20 compares the voltammograms recorded in the steady state and under conditions of non-steady state diffusion and also illustrates the concentration profiles for R and O in both situations. As discussed earlier, in the steady state, there is a fixed diffusion layer and the change in current density arises only from the change in the surface concentration of R as the potential is scanned positive; the surface concentration decreases with a consequent increase in the flux of R to the surface (and current density) until the surface concentration reaches zero, when both reach a constant value. In the non-steady state, the potential is changed much more rapidly so that the concentration profile cannot reach the steady state thickness. As a result two changes occur simultaneously; as the surface concentration is decreasing, non-steady state diffusion is occurring close to the surface (*cf.* a potential step experiment, Figure 7.13), seeking to minimize differences in concentration and causing the concentration profile to expand into solution. Clearly, at all potentials in the rising portion of the voltammogram, the flux of R and the current density are higher in the non-steady state experiment. In the non-steady state, after the surface concentration reaches zero, the concentration profile can only expand into solution (*cf.* a potential step, Figure 7.13) and then the flux

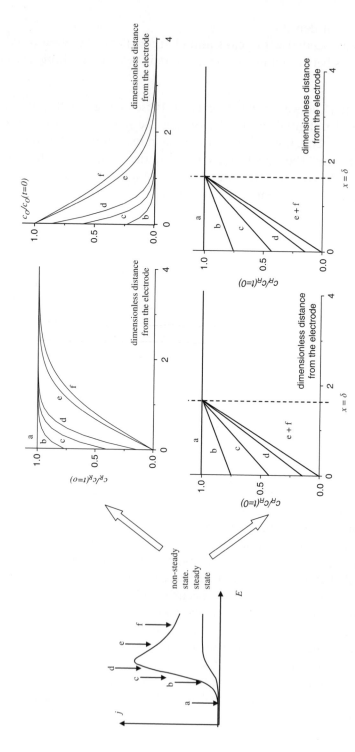

Figure 7.20 Understanding the shape of a cyclic voltammogram. Comparison of the concentration profiles under conditions of steady state mass transport and non-steady state diffusion. The profiles are labelled with the potentials shown on the voltammograms.

of R and the current density will pass through a maximum value and then decrease – a peak in the current will be seen. The influence of potential scan rate is easily understood qualitatively. With an increase in potential scan rate, the time to change the surface concentration will decrease and hence the diffusion layer will always be thinner. The current density at all potentials, including the peak potential, will be larger. The overall shape of the peak should be stressed; the current density rises steeply but beyond the peak it decays relatively slowly towards the steady state limiting current density. The steepness of the rising part of the voltammogram and the value of the peak current density (as well as its dependence on potential scan rate) require a complete theoretical description and mathematical solution (see below).

In a cyclic voltammogram, the reverse scan is also of interest. While the potential remains well positive to the equilibrium potential for the couple O/R, the surface concentration of R is not affected by the direction of scan and it will remain zero – the current density will continue to decay towards the steady state oxidation limiting current density. As the potential approaches the formal potential for the couple O/R on the reverse scan, R as well as O will again be needed at the electrode surface (*e.g.* for a reversible reaction, to meet the demands of the Nernst equation) and the only way for R to be created rapidly is for reduction of O close to the surface to occur. Thus the current density will cross the current density axis and change sign. The response on the reverse scan will also pass through a peak because of non-steady state diffusion with respect to O. Figure 7.21 shows the concentration profiles for both O and R during this part of the potential scan. The shape becomes more complex but can be understood in terms of the changes in

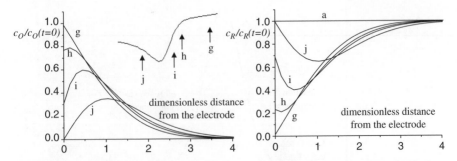

Figure 7.21 Concentration profiles for both O and the reactant, R, during the back scan of a cyclic voltammogram for the oxidation of R to O. The potential for each profile is shown on the inset, the voltammetric response during the reverse scan. The profile 'a' corresponds to the initial potential for the cyclic voltammogram.

concentrations at the electrode surface and continued expansion of the diffusion layer into the bulk solution.

When the electrode reaction under study leads to the formation of a monolayer, or involves oxidation/reduction of a layer on the surface, the voltammetric peak has a quite different shape. This results because of the limited amount of reactant and/or sites on the electrode surface. As the scan approaches the formal potential for the reaction, the current density will again increase sharply. But as the electrode reaction occurs, a limited amount of reactant is being consumed (or surface sites filled) so the current density must pass through a peak and, indeed, the current density must decay to zero when all the reactant is consumed. Hence, the shape of peak observed for such a reaction will be quite different from that resulting from non-steady state diffusion. It will be much more symmetrical with the current density being zero both positive and negative to the peak (Figure 7.22). The charge density associated with the reaction (the area under the peak) can also be interpreted; for a process limited to a monolayer, the charge density will be in the range 100–200 $\mu C\,cm^{-2}$. In addition, this charge density will be determined only by the 'amount of reactant' and hence will be independent of potential scan rate. This leads to a peak current density that is proportional to scan rate.

7.3.2.2 Reversible Electron Transfer.

The simplest case to consider is an electron-transfer reaction:

$$R - ne^- \rightleftarrows O \qquad (7.47)$$

where both O and R are dissolved and stable in the electrolyte and electron transfer is rapid. The qualitative form of the response has

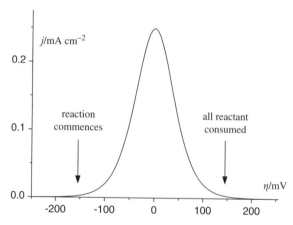

Figure 7.22 Cyclic voltammogram for the oxidation of a monolayer of a surface-bound species when the electron-transfer reaction is rapid.

already been discussed in the previous section. The exact form of the voltammogram and its dependence on potential scan rate results from the solution of the equations:

$$\frac{\partial c_R(x, t)}{\partial t} = D\frac{\partial^2 c_R(x, t)}{\partial x^2} \quad \text{and} \quad \frac{\partial c_O(x, t)}{\partial t} = D\frac{\partial^2 c_O(x, t)}{\partial x^2} \tag{7.48}$$

In general, cyclic voltammetry is carried out with only the oxidized or the reduced species in solution; here the case where the initial solution contains only R will be considered (but the conclusions would be identical for a solution containing only O if the direction of the forward scan were reversed – a cathodic current density would then be observed). For this experiment, the initial conditions are at $t = 0$:

$$c_R(x, t) = c_R \quad \text{and} \quad c_O(x, t) = 0 \tag{7.49}$$

and the boundary conditions, at $x = \infty$: $c_R(x,t) = c_R$ and $c_O(x,t) = 0$; and at $x = 0$:

$$D\left(\frac{\partial c_R}{\partial x}\right)_{x=0} = -D\left(\frac{\partial c_O}{\partial x}\right)_{x=0} \tag{7.50}$$

and:

$$\frac{c_O(0, t)}{c_R(0, t)} = \exp\frac{nF\left(E_1 + vt - E_e^\circ\right)}{RT} \tag{7.51}$$

Equation (7.51) is the Nernst equation written with the ratio of the concentrations as the subject and recognizing that, since the potential at any time during the experiment depends on the potential at the start of the scan and the scan rate, $E - E_e^\circ = E_1 + vt - E_e^\circ$. The solution of this set of equations is complex but possible. It confirms that, for a reversible couple, only thermodynamic information may be extracted. The formal potential is found from:

$$E_e^\circ = \frac{E_p^a + E_p^c}{2} \tag{7.52}$$

and the diffusion coefficient from the Randles–Sevèik equation (at 298 K):

$$j_p = 2.69 \times 10^5 \, n^{3/2} D^{1/2} c v^{1/2} \tag{7.53}$$

Qualitatively and with experience, the shape of the voltammogram for a reversible couple can be judged by eye. Quantitatively, the decision can be made using the well-known tests in Table 7.1 (the symbols are defined in Figure 7.23). These tests, however, place heavy emphasis on a few

Table 7.1 Diagnostic tests for the form of the cyclic voltammetric responses for a reversible and irreversible couple O/R at 298 K.

Experimental measurable	Reversible couple	Irreversible couple
$\Delta E_p = E_p^a - E_p^c$	$\dfrac{59}{n}$ mV	$> \dfrac{59}{n}$ mV and $f(k_s, \nu)$
$E_p - E_{p/2}$	$\dfrac{59}{n}$ mV	$\dfrac{48}{\alpha n}$ mV
$\left\| \dfrac{j_p^c}{j_p^a} \right\|$	1	1
E_p	Independent of ν	Dependent on ν
j_p	Proportional to $\nu^{1/2}$	Proportional to $\nu^{1/2}$

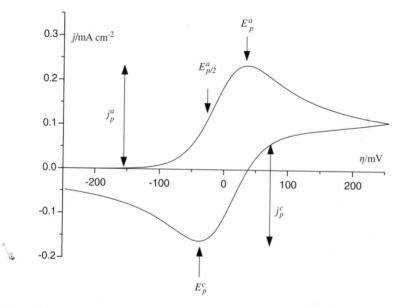

Figure 7.23 Cyclic voltammogram to define the symbols used in the discussion of the shape of the response. In fact, the voltammogram is for the oxidation of R (1 mM) with a potential scan rate 100 mV s^{-1}; $D_R = 10^{-5}$ cm^2 s^{-1}.

particular points on the responses (*i.e.* the peak and half-peak positions) and it is better to compare the shape of the whole voltammogram with a simulated response. Notably, the shape of the voltammogram, but not the magnitudes of the current densities, is independent of the potential

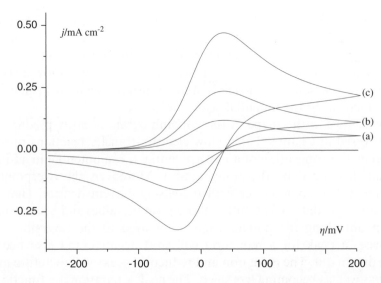

Figure 7.24 Influence of potential scan rate of the cyclic voltammogram; the potential scan rates are (a) 25 (b) 100 and (c) 400 mV s^{-1}. The voltammograms are for the oxidation of R (1 mM); $D_R = 10^{-5}$ cm^2 s^{-1}.

scan rate. The current densities increase with potential scan rate (Figure 7.24) because, as the scan rate increases, the surface concentration of reactant changes more rapidly – less diffusion can occur during the timescale of the potential scan and the flux of reactant to the surface at each potential will be higher. In fact, the peak current density is proportional to the square root of the potential scan rate, see Equation (7.53). As a result, if they are presented as plots of $j/v^{1/2}$ vs E the voltammograms at all scan rates will collapse onto a single response. This is a useful trick for identifying changes with potential scan rate from those associated with non-steady state diffusion. A further feature should be noted. The charge (area under the response) associated with the forward reaction, here R $\rightarrow$ O, is large compared to the charge associated with the back reaction, here O $\rightarrow$ R. This must be the case, since throughout the experiment there is a concentration gradient driving the product O into the bulk solution (Figures 7.20 and 7.21).

7.3.2.3 Irreversible Electron Transfer. The voltammogram has the shape associated with a reversible couple O/R provided the kinetics of the electron-transfer reaction are fast enough that the surface concentrations of O and R are those predicted by thermodynamics under the prevailing mass transfer regime. On the other hand, when the kinetics of electron transfer are insufficient to maintain equilibrium at the surface, the reaction becomes irreversible and the theoretical description of the experiment must include the surface kinetics. This may be necessary if (a) the standard rate constant drops below a critical value or (b) the rate of non-steady state diffusion is increased sufficiently by an increase in the potential scan rate.

The change in the shape of the voltammogram is easily predicted. It was stressed in Chapter 1 that slow electron-transfer reactions must be driven by the application of an overpotential (*i.e.* a potential in addition to that demanded by thermodynamics). Moreover, the overpotential required will become larger with increasing current density. Hence, a peak for an oxidation will shift to more positive values and become more drawn out along the potential axis compared to the reversible case. Likewise, a peak for a reduction will shift negative and also become more drawn out. The oxidation and reduction peaks will therefore move apart as well as becoming less steep. The peak separation is a function of both the standard rate constant and the potential scan rate. Beyond the peaks, however, the reactions still become diffusion controlled and the current density decays towards the steady state value. Peaks for irreversible processes are slightly smaller than those for reversible processes; they are less steep and it takes a longer time for the surface concentration of reactant to be driven down to zero, reducing the flux of reactant at the peak potential. Figure 7.25 compares the voltammograms for the reversible and irreversible cases while Figure 7.26 illustrates the influence of the potential scan rate; in the latter figure, $j/\nu^{1/2}$ is plotted against E to allow the presentation without the influence of non-steady state diffusion leading to large differences in the current density scale. Table 7.1 summarizes the characteristics for an irreversible electron-transfer reaction. Notably, while the peak current density is proportional to the square root of the potential scan rate, the slope of a j_p vs $\nu^{1/2}$ plot is not the same as for the reversible case. The expression for the irreversible case is:

$$j_p = 3 \times 10^5 \, n(n\alpha)^{1/2} D^{1/2} c \nu^{1/2} \tag{7.54}$$

The kinetics of the electron-transfer reaction may be obtained from the voltammogram. The transfer coefficient, α, may be obtained from the peak shape (Table 7.1). The standard rate constant can be determined from the peak separation as a function of the potential scan rate using a plot such as that shown in Figure 7.27.

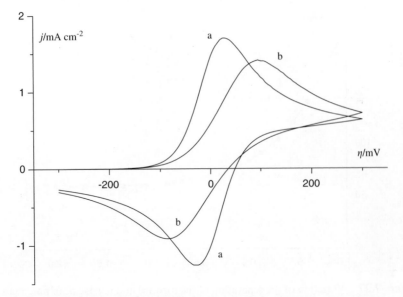

Figure 7.25   Cyclic voltammograms for (a) a reversible and (b) an irreversible couple
O/R.

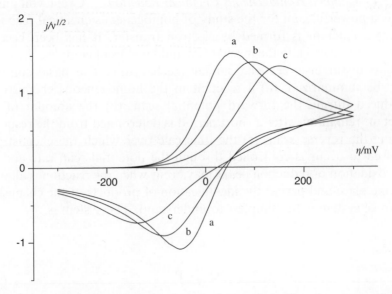

Figure 7.26   Influence of the potential scan rate on the cyclic voltammetry of an irreversible couple O/R: (a) ν, (b) $10\,\nu$ and (c) $100\,\nu$. To avoid changes in the
current density scale due to differences in the rate of non-steady state
diffusion, the voltammograms are presented as plots of $j/\nu^{1/2}$ *vs* η.

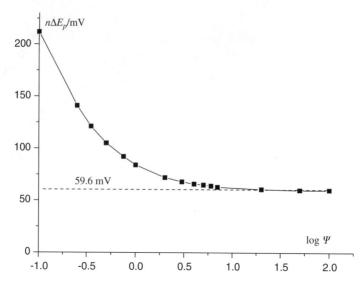

Figure 7.27   Variation of the separation of the forward and reverse peak potentials on a cyclic voltammogram for a quasi-reversible couple O/R with the standard rate constant for the couple and the potential scan rate; $\alpha = 0.5$. $\Psi = k_s(RT)^{1/2}/(\pi D\nu n F)^{1/2}$. Data taken from R. S. Nicholson, *Anal. Chem.*, 1965, **37**, 1351.

7.3.2.4   Coupled Homogeneous Chemical Reactions.

Cyclic voltammetry is a powerful tool for the study of homogeneous chemical reactions where a reactant is formed by electron transfer. It has been become possible to investigate the mechanism and kinetics of complex chemistry and to distinguish between similar mechanisms. The basic concepts may be summarized: (a) a reactant in the homogeneous chemistry is produced during the forward potential scan, (b) the amount of the reactant remaining after a short period is determined from the response during the reverse scan, (c) the timescale over which the chemistry is allowed to occur is controlled through the potential scan rate and (d) new oxidation or reduction peaks may occur when the reaction is allowed to take place and permit the identification of products of the chemistry.

An ec system is the simplest to understand. The system is:

$$R - e^- \rightleftharpoons O \tag{7.55}$$

$$O \xrightarrow{k} P \tag{7.56}$$

where the electron-transfer reaction is reversible and the intermediate, P, has no electrochemistry. The form of the voltammogram will depend on the rate constant, k, for the chemical step (7.56) and the potential scan rate, ν. Initially, the discussion will assume that the rate constant has a particular value and consider the influence of potential scan rate. Three situations can arise, depending on the relative values of the half-life of the intermediate, O $(\tau_{1/2} = (\log 2)/k)$, and the timescale of the experiment determined by the potential scan rate:

1. When the timescale of the experiment is short compared to the half-life of the intermediate, O, the response cannot be influenced significantly by the chemical reaction and the voltammogram will be that for a reversible 1e$^-$ reaction (curve (c) of Figure 7.28).
2. In contrast, when the potential scan rate is decreased sufficiently so that the timescale of the experiment is long compared to the half-life, effectively all of the species O will react to give P and there can be no peak on the reverse scan for the reduction, O→R. The reverse scan does show a wave, but this corresponds only to the generation of O switching off as the potential comes into the range where reaction R→O is no longer favourable [see curve (a) of Figure 7.28]. The form of the forward scan also shows slight changes. The oxidation peak shows a small shift to less positive potentials, *i.e.* the electrode reaction occurs more easily. There are two ways to view this shift: (a) the thermodynamics should now

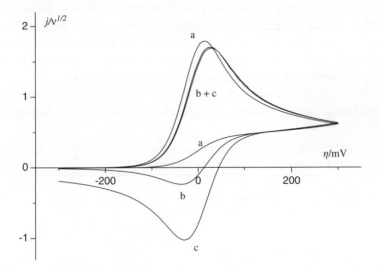

Figure 7.28   Influence of the potential scan rate on the cyclic voltammogram for an ec system, Equations (7.55) and (7.56): (a) 0.1ν, (b) ν and (c) 10ν. To avoid changes in the current density scale due to differences in the rate of non-steady state diffusion, the voltammograms are presented as plots of $j/\nu^{1/2}$ vs η.

consider the conversion R→P rather than R→O and, since the reaction O→P is very favourable, it leads to a decrease in the positive Gibbs free energy for the overall reaction; (b) at a point in the rising part of the voltammogram, the ratio of surface concentrations, $(c_O/c_R)_{x=0}$, is determined by the Nernst equation. The fact that the chemical reaction O→P is taking place means that a larger current density must be passed to maintain this ratio. Hence, at all potentials in this range, the current density will be higher and this corresponds to a shift in the peak.

3. At intermediate scan rates, when the timescale of the experiment and the half-life of the intermediate are similar, the voltammogram will have the form shown in Figure 7.28(b). Some reverse peak will be observed but it will be smaller than in the absence of a chemical reaction. The ratio of the peak current densities, $|j_p^C/j_p^A|$ will increase with increasing potential scan rate. Also, there will be a trend for the forward peak to shift to less positive potentials with decreasing potential scan rates.

The trend in voltammograms from curve (a) to curve (c) will also be seen in experiments where the potential scan rate is maintained constant and the rate constant for the homogeneous chemical reaction is decreased (*e.g.* by lowering the temperature). Also, when studying a series of compounds (*e.g.* aromatic molecules with different substituents), the trend in the shapes of the voltammograms might be seen, as the rate of the coupled chemistry varies. In this case, however, there will also be potential shifts due to a dependence of the formal potential on the electron-donating/withdrawing effect of the substituents.

Table 7.2 lists some important characteristics suitable for the recognition of an ec reaction. As with all mechanisms, it is important to

Table 7.2 Characteristics of a cyclic voltammogram for an ec reaction where the electrode reaction is an oxidation.

$\left|\dfrac{j_p^C}{j_p^A}\right| < 1$ but increases with increasing ν

$\dfrac{j_p^A}{\nu^{1/2}}$ decreases slightly with decreasing ν

E_p^A is negative to E_e^O and shifts negative with decreasing ν

When the reverse peak is not observed $\dfrac{dE_p^A}{d\log\nu} = -28\text{mV}$

recognize that all the features of the cyclic voltammogram change simultaneously as the conditions are varied and the response goes from that for a reversible to that for an irreversible system. The rate constant for an ec reaction can therefore be determined from the variation of the ratio of the peak current densities or the shift in the peak potential with potential scan rate. Appropriate plots are shown in Figure 7.29; both have a linear region corresponding to the couple O/R being fully reversible and another where the homogeneous chemical reaction is having a large influence on the response. The rate constant is conveniently determined from the intersection of the two linear portions. Alternatively, and perhaps better, the rate constant can be determined by comparing the experimental responses as a function of potential scan rate with simulated voltammograms as a function of scan rate and rate constant. As stressed previously, determination of the rate constant depends on matching the half-life of O to the timescale of the experiment; with practical potential scan rates, this means that rate constants in the range 10^{-2}–$10^2\,\mathrm{s}^{-1}$ can be estimated by cyclic voltammetry.

Very commonly, the product of the homogeneous chemical reaction, P, will be electroactive (in particular, the oxidation or reduction of organic molecules almost always involve $2e^-$; radicals and ion radicals are seldom stable and the cleavage/formation of a bond is also a $2e^-$ process). The further electron transfer may occur at the same potential as the oxidation of R when an increase in peak height will occur at either

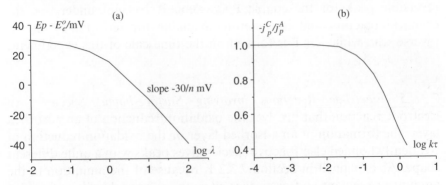

Figure 7.29 Dimensionless plots for determination of the rate constant for the homogeneous chemical reaction in an ec mechanism, Reactions (7.55) and (7.56). (*a*) Dependence of the potential for the oxidation (R to O) peak on the rate constant and the potential scan rate $\lambda = (k/v)(RT/nF)$ and (*b*) the variation of the peak current ratio, $-j_p^C/j_p^A$, on the rate constant and the potential scan rate, τ is the time taken for the scan from the formal potential for the couple R/O to the potential where the direction of scan is reversed (hence a function of v). Data taken from R. S. Nicholson and I. Shain, *Anal. Chem.*, 1964, **36**, 706.

more positive or negative potentials when new peaks will result from the homogeneous chemistry. The new peaks may be either anodic or cathodic. It is then that variation of potential limits, scan rates and changes with cycling become informative.

As an illustration, consider an ece system:

$$O + 2e^- \rightleftharpoons R \qquad\qquad (7.57)$$

$$R \overset{k}{\rightleftharpoons} P \qquad\qquad (7.58)$$

$$P + 2e^- \rightleftharpoons Q \qquad\qquad (7.59)$$

where P is reduced at a potential negative to that for the reduction of O. The limiting situations are clear. If the chemistry is slow (on the time-scale of the experiment), the voltammogram will show only a single, reversible reduction process for the couple O/R (Figure 7.30a). Note the voltammograms in the figure are all presented as plots of $j/\nu^{1/2}$ *vs* E to avoid the influence of non-steady state diffusion on the current densities. In contrast, if the chemistry is fast, the voltammogram will show two reduction processes (Figure 7.30c). The response for the reduction of O will be that for an irreversible $2e^-$ reaction and a further reversible $2e^-$ process for the reaction will be seen at more negative potentials. The two reduction peaks will have a similar height since both involve two elec-trons. At intermediate scan rates, the voltammogram (Figure 7.30b) will show a partially reversible response for the couple O/R and a smaller reversible peak for the couple P/Q; some R has not undergone the chemical reaction and is therefore available for reoxidation on the reverse scan while less P is formed on the timescale of this experiment.

7.3.2.5 Electrode Reactions involving Surface-bound Species.

An electrode reaction that involves the oxidation/reduction of an adsorbed layer, the formation of an adsorbed layer or the oxidation/reduction of covalently bonded electroactive species gives peaks with a quite different shape. As explained in Section 7.3.2.1, because of the limitation in the amount of reactant and/or surface sites, the current density will return to zero at potentials beyond the peak. If the electron-transfer reaction is rapid, *e.g.*:

$$R_{ADS} - e^- \rightleftharpoons O_{ADS} \qquad\qquad (7.60)$$

the voltammogram will have the form illustrated in Figure 7.31 and the characteristics listed in Table 7.3. Notable features are the symmetrical

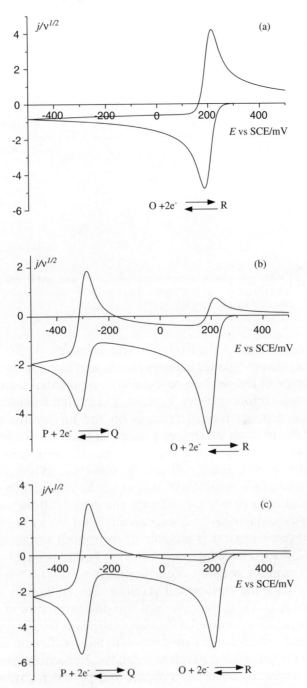

Figure 7.30 Cyclic voltammograms for an ece reaction represented by Reactions (7.57–7.59) where the formal potential for the couple P/Q is negative to that for the couple O/R but both $2e^-$ couples are reversible. The potential scan rates are (a) $100\,\nu$, (b) ν and (c) $0.01\,\nu$.

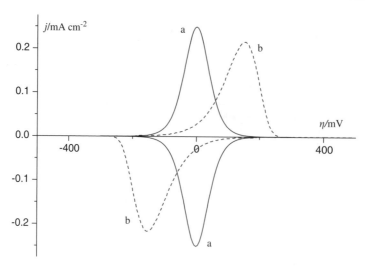

Figure 7.31 Cyclic voltammograms for the oxidation and reduction of a surface-bound species. (*a*) Reversible (fast) electron transfer and (*b*) irreversible (slow) electron transfer.

shape, the occurrence of the oxidation and reduction peaks at the same potential, the charge balance between anodic and cathodic processes and the dependence of the peak current density on potential scan rate. Such processes are therefore easily recognized. The charge balance arises only when all the product formed remains on the surface for the reverse reaction. If the product undergoes a chemical reaction or the product is soluble in the electrolyte, the size or shape of the reverse peak will change and charge balance will not be observed. When the electron transfer is slow, an overpotential will again be required to drive both oxidation and reduction steps and then the peaks will become broader and move apart; the other characteristics remain the same (Table 7.3).

In some experiments, it is possible to form both surface bound and dissolved product. The formation of surface bound product will lead to the electron-transfer reaction occurring more readily as the formation of the bond with the surface will stabilize the product; in the case of oxidation of R to O, the anodic peak for the formation of O_{ADS} will occur at less positive potential. Of course, the peaks for the formation of adsorbed and solution free product will have different shapes and dependence on potential scan rates, see above. In consequence, the peak for the formation of adsorbed product will be most obvious at high potential scan rates as well as with lower concentrations of reactant. Similarly, if the reactant is adsorbed, separate peaks for the oxidation/reduction of the adsorbed and solution free reactant will again be

Table 7.3 Diagnostic tests for the form of the cyclic voltammetric responses for a reversible and irreversible couple O/R when O and/or R are surface bound at 298 K.

Experimental measurable	Reversible couple	Irreversible couple
$\Delta E_p = E_p^a - E_p^c$	$0\,mV$	$>0\,mV$ and $f(k_s, v)$
$\left\| \dfrac{j_p^c}{j_p^a} \right\|$	1	1
E_p	Independent of v	Function of v
j_p	Proportional to v	Proportional to v
q_p	$q_p^a = q_p^c$	$q_p^a = q_p^c$
Peak shape	Symmetrical	Symmetrical but broader

observed but the oxidation/reduction of the adsorbed reactant will be more difficult.

7.3.2.6 Electrode Reactions Involving Phase Formation.

Figure 7.32 shows a typical cyclic voltammogram for a M^{n+}/M couple at a cathode other than M (commonly carbon) and with a solution containing a relatively low concentration of M^{n+}. Once again the overall shape of the voltammogram is significantly different:

1. The cathodic peak on the forward scan has a similar shape to those discussed above for the reduction of a dissolved reactant where the reaction becomes diffusion controlled beyond the peak. In fact, detailed analysis of the foot of the wave would show that the current density rises unusually steeply.
2. On the reverse scan, there is a potential range where the cathodic current is higher on the reverse scan than on the forward scan (also seen in the RDE experiment, Figure 7.10). This feature arises because of the need to form nuclei of the metal phase on the foreign substrate. Formation of stable nuclei is always a difficult process because small nuclei with a high surface area and little volume tend to redissolve. Hence, creation of the metal phase requires an overpotential. On the reverse scan, the metal phase already exists on the electrode and metal deposition can continue until the reduction $M^{n+} \rightarrow M$ is no longer favourable. In other words, in the region of the nucleation loop, during the forward scan one is trying

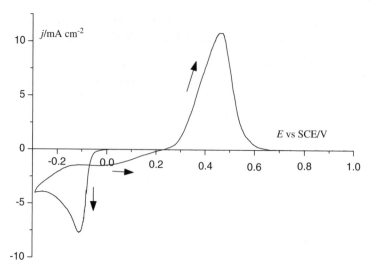

Figure 7.32 Cyclic voltammogram for the deposition and dissolution of palladium at a vitreous carbon disc electrode. Solution: 10 mM $PdCl_2$ + 1 M KCl, pH 1. Potential scan rate: 100 mV s^{-1}.

but failing to form the nuclei of the metal phase, while on the reverse scan, deposit thickening can occur readily. The steep rise in the foot of the cathodic peak also results from the need for nucleation; as the stable nuclei are created, there is already an overpotential for the reduction of M^{n+}. In addition, the surface area of M available for the cathodic reduction is expanding rapidly.

3. The reverse scan shows a sharp, symmetrical anodic peak for the oxidation of the metal layer back to M^{n+} in solution (*i.e.* a stripping peak). The peak is symmetrical because the charge for metal dissolution is limited by the amount of metal plated. Indeed, the charge associated with metal dissolution must be equal to the *total* cathodic charge (note that metal is deposited during both forward and reverse scans) provided that there are no competing electrode reactions. The current density for metal dissolution can be large because the metal is on the surface and no mass transport process limits the rate of the metal dissolution. If the kinetics of the M/M^{n+} couple are rapid, as for the Pd(II)/Pd couple in Figure 7.32, the current density on the reverse scan will pass steeply through the zero current axis. The potential at the crossover will be a good estimate of the equilibrium potential for the M/M^{n+} couple and the slope of the voltammogram through the zero current axis reflects its kinetics.

7.3.2.7 The Approach to the Study of a New System. Typically, the experimental approach to the study of a system whose chemistry is unknown will have the following stages:

1. Unless there is a particular reason, all cyclic voltammograms should have an initial potential, E_1, where the current density is zero. Hence, the first step is to find experimentally the potential range where the current density is at the background level. This can be achieved either by applying a series of potentials and noting the current density or by running a few preliminary voltammograms.
2. Having selected E_1, record a series of voltammograms at a convenient scan rate (*e.g.* $100 \, \text{mV s}^{-1}$) where the reversal potential, E_2, is extended stepwise. The number of anodic and cathodic peaks and their general characteristics (shape *etc.*) should be noted and variation of the scan rate considered at selected values of E_2. Particular attention should be paid to conditions where the importance of a peak depends on the potential limits and/or potential scan rate. These data are then used to form preliminary ideas about the chemistry occurring.
3. The potential limits should then be reset so that only the primary electrode process is seen on the forward scan and the potential scan rate varied so as to make a more quantitative assessment of the characteristics of the voltammogram (using peak potential peak current density, charge balance, *etc.*) as discussed in the previous sections.
4. As the primary process is better understood, again extend the potential limits for a more detailed study of the further peaks revealed.
5. Consider additional experiments (*e.g.* comparing 1st and n^{th} cycles, holding the potential at selected values prior to a scan, modifying the pH, varying reactant concentration, temperature) to provide tests for the mechanisms being considered.

As emphasized previously, such a set of experiments need take only a couple of hours but it important to analyse data as the experiments are being carried out. The full benefit cannot be obtained by carrying out a set of predetermined experiments and analysing the data at a later date. Later experiments will seek to confirm the mechanism and obtain kinetic parameters. A digital simulation that gives a good fit to the experimental data is a satisfying conclusion to a study. Hints as to approaches to improving the quality of the experimental data and avoiding unwanted distortion of the responses were set out in Chapter 6.

The thinking underlying cyclic voltammetric investigations can be illustrated by looking at the voltammetry for the reduction of nitrobenzene in various media; in all cases, the electrode is a vitreous carbon disc and a suitable concentration of nitrobenzene would be 1–5 mM. Figure 7.33 shows some voltammograms for nitrobenzene in (a) dimethylformamide (DMF) + 0.2 M Bu_4NBF_4, (b) aqueous 1 M H_2SO_4 and (c) aqueous acetate buffer, pH 4.2.

Following the procedure outlined above, in the aprotic medium, dimethylformamide + 0.2 M Bu_4NBF_4, the primary reduction process was defined as occurring just negative to −1.0 V *vs* SCE. Hence, the potential limits were set to + 100 and −1250 mV, where only a single cathodic peak and coupled anodic peak were observed (Figure 7.33a). Then the potential scan rate was varied. The response has all the expected characteristics for a reversible 1e$^-$ process, see Table 7.1, and it can be concluded that the electrode reaction leads to the stable nitrobenzene anion radical:

$$\tag{7.61}$$

Extending the negative potential limit leads to several further cathodic peaks; they are all irreversible and their peak current densities do not relate simply to that for the primary cathodic process. The chemistry in this potential range is complex and preparative electrolyses reveal several products arising from dimerization, cleavage of the nitro group from the benzene ring and protonation. Adding an excess of acetic acid to the aprotic medium changes the voltammetry significantly. The primary reduction peak moves positive to *ca.* −0.8 V and the peak becomes much larger; comparison of the peak height with that before the addition of acetic acid suggest the transfer of ∼4e$^-$. Clearly, in the presence of proton donor, the anion radical is protonated and this leads to a cascade of chemistry and further electron transfers. The large positive shift of peak potentials indicates that the coupled homogeneous chemistry is very favourable.

When the cyclic voltammetry is repeated in aqueous acid (Figure 7.33b) the response is very simple. A single cathodic peak is seen at −740 mV *vs* SCE and at potentials negative to −1.0 V only a very large current density leading to H_2 gas evolution is observed. At negative potentials, there are no anodic peaks on the reverse scan. The cathodic peak is always totally irreversible but the peak current density is proportional to the square root of the potential scan rate, confirming that the reduction becomes diffusion controlled. Estimates of the peak

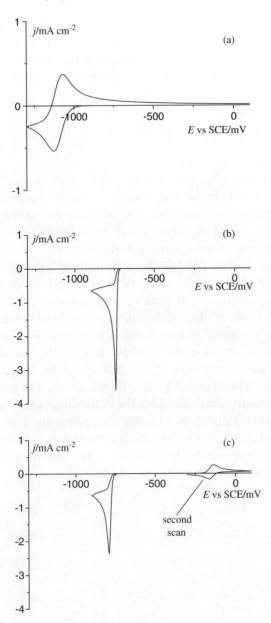

Figure 7.33 Cyclic voltammograms at a vitreous carbon disc electrode for 2 mM nitrobenzene in (*a*) DMF/0.1 M Bu$_4$NBF$_4$, (*b*) 1 M aqueous H$_2$SO$_4$ and (*c*) aqueous acetate buffer, pH 4.2. Potential scan rate 100 mV s^{-1}.

current densities using a guessed value of the diffusion coefficient for nitrobenzene suggests that n is large and the transfer of $6e^-$ would not be unreasonable. This would be consistent with chemical reductions in strong aqueous acid, where aniline is the usual product:

$$\text{NO}_2\text{-C}_6\text{H}_5 + 4\,\text{H}^+ + 6e^- \longrightarrow \text{NH}_2\text{-C}_6\text{H}_5 + 2\text{H}_2\text{O}$$

(7.62)

Indeed, if the potential is extended to $+0.7\,$V and the potential is cycled for some time, small anodic and coupled cathodic peaks start to appear at positive potentials; by comparison with the literature or examining the voltammetry of aniline, these could be assigned to the formation of some poly(aniline) on the electrode surface.

In the acetate buffer, pH 4.2, the voltammetry again reveals a single, irreversible cathodic peak at $-780\,$mV vs SCE with H_2 evolution seen at more negative potentials. The peak current density is again proportional to the square root of the potential scan rate but smaller than that observed in the stronger acid medium. If the value of $6e^-$ in sulfuric acid were accepted, it would be concluded that, in the acetate buffer, the reduction of nitrobenzene is a $4e^-$ process. The cyclic voltammetry with a positive limit of $+0.1\,$V also shows an anodic peak at $-0.15\,$V provided the negative limit includes the cathodic peak. The reduction of nitrobenzene under these conditions is leading to a product that is readily oxidized. Also, if a second cycle is recorded, a new cathodic peak is seen and this appears to be coupled to the anodic peak. Chemistry consistent with this voltammetry would be:

$$\text{NO}_2\text{-C}_6\text{H}_5 + 4\,\text{H}^+ + 4e^- \longrightarrow \text{NHOH-C}_6\text{H}_5 + \text{H}_2\text{O}$$

(7.63)

$$\text{NHOH-C}_6\text{H}_5 - 2e^- \rightleftarrows \text{NO-C}_6\text{H}_5 + 2\text{H}^+$$

(7.64)

This theory could be tested by investigating the electrochemistry of phenylhydroxylamine or nitrosobenzene in the same acetate buffer. Both compounds would show a reversible $2e^-$ process at $-0.15\,$V. Alternatively, the mechanism could be confirmed by preparative electrolysis to obtain coulometric values for n and also to identify the products.

7.3.3 AC Impedance

In an AC impedance experiment a small amplitude, sinusoidal potential:

$$E = \Delta E \sin 2\pi ft \tag{7.65}$$

is superimposed upon an applied potential, commonly the equilibrium potential, and the current response is monitored. For most electrode reactions, the current response will also be sinusoidal, with the same frequency, f, but will be different in amplitude and phase from the applied potential, *i.e.*:

$$I = \Delta I \sin (2\pi ft + \phi) \tag{7.66}$$

In these equations, ΔE is the maximum amplitude of the applied sinusoidal potential [usually selected to be $<10\,\text{mV}$ so that the linear approximation to the Butler–Volmer equation, (1.48), can be used to describe the kinetics of electron transfer], ΔI is the maximum amplitude of the current response and ϕ is the phase angle between the current output and the potential perturbation. The phase shift arises because electrode reactions usually behave equivalently to an electrical circuit that contains capacitances as well as resistances.

In the most common experiment, the impedance of the electrode, Z:

$$Z = \frac{E}{I} \tag{7.67}$$

is analysed as a function of the frequency of the potential perturbation, f, and in view of the phase shift in the current response, see Equation (7.66), Z is usually a complex quantity. Commonly, the frequency range investigated is 10^{-2}–$10^{4}\,\text{s}^{-1}$. This is a powerful experiment because the single experiment can contain information about a very wide range of timescales (*e.g.* relative rates of the kinetics of the electrode reaction and non-steady state diffusion). However, it is often difficult to present the results in a form that is easily interpreted qualitatively. Furthermore, quantitative analysis normally requires the construction of an equivalent circuit, *i.e.* an electrical circuit that mimics the behaviour of the electrode reaction, and the data are extracted by matching the electrical responses of the electrode reaction and the equivalent circuit. Even for the simplest electrode reaction, the equivalent circuit has several components, see Figure 7.34, and many electrode reactions require significantly more complex equivalent circuits.

One approach to the interpretation of AC impedance data is the use of the Argand diagram, where the imaginary component of the impedance is plotted *versus* the real component as a function of frequency. The real component is the in-phase component (it varies with $\sin 2\pi ft$) while the imaginary component is shifted by $\pi/2$ [it varies with $\sin(2\pi ft + \pi/2)$]. With modern instrumentation these components are determined using phase-sensitive detection or Fourier-transform analysis and are presented as an Argand diagram.

Figure 7.34 shows an equivalent circuit for the simple redox reaction:

$$O + ne^- \rightleftharpoons R \qquad (7.68)$$

and the Argand diagram that would arise for this equivalent circuit. Note that the equivalent circuit recognizes that the electrode will behave as a circuit with both capacitative and resistive properties. In fact, the circuit has a charge transfer resistance and a Warburg impedance in parallel with the double layer capacitance and also recognizes that the experimental cell will have an uncompensated resistance. The Warburg impedance is largely a resistance associated with the diffusion processes but is also a pseudo-capacitance associated with electron transfer.

(a)

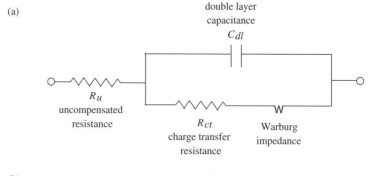

(b)

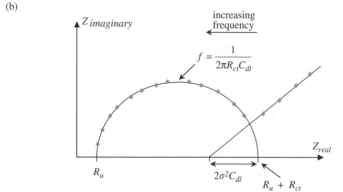

Figure 7.34   (*a*) Equivalent circuit for a simple electron-transfer couple O/R at an inert electrode held at the equilibrium potential for a solution containing O and R. (*b*) The corresponding Argand diagram.

Kinetic parameters would be obtained from an AC experiment carried out around the equilibrium potential for an inert electrode in a still solution containing both O and R. The experimental data, presented as an Argand diagram, would allow values to be obtained for each of the circuit components. These must then be related to the parameters associated with the electrode reaction, recognizing that, because the experiment covers a very wide range of frequencies and hence timescales, they could contain information about both electron transfer and diffusion. Interpretation of the circuit components requires a theoretical description and a mathematical solution; in fact, Fick's second law for O and R with the appropriate initial and boundary experiments. For the kinetics of electron transfer, R_{ct} is taken from the diameter of the semi-circle at high frequencies and the theoretical description of the experiment would show that:

$$R_{ct} = \frac{RT}{nF} \frac{1}{j_o} \tag{7.69}$$

At low frequencies, the diffusion processes predominate and this leads to a linear region of the Argand diagram with unit slope. The diffusion coefficient can be obtained from the intercept on the real axis since:

$$\sigma = \frac{RT}{2^{1/2} n^2 F^2 A D^{1/2}} \left(\frac{1}{c_O} + \frac{1}{c_R} \right) \tag{7.70}$$

and the double layer capacitance, C_{dl}, can be found from the frequency at the maximum of the high frequency semi-circle:

$$f = \frac{1}{2\pi R_{ct} C_{dl}} \tag{7.71}$$

To obtain the complete Argand diagram shown in Figure 7.34, it must be possible to obtain data over a wide and appropriate frequency range, *i.e.* frequencies where electron transfer and diffusion dominate, and this will depend on the kinetics of electron transfer. In some systems, it will be possible to observe only the semi-circle or only the linear regions.

In systems where the charge transfer resistance is very high, *i.e.* an ideally polarized electrode (Chapter 3) such as the mercury/aqueous NaF interface, the impedance is purely capacitative. The current output will be out of phase with the potential perturbation by $\pi/2$ and the resulting imaginary impedance will be given by:

$$Z_{imaginary} = \frac{1}{2\pi f C_{dl}} \tag{7.72}$$

It is found experimentally as the ratio of the amplitudes of the AC current to the potential perturbation. This provides a very simple way to measure the double layer capacitance. When the electrode/electrolyte interface is not perfectly ideal, the double layer capacitance can still be estimated by using a high frequency and determining the out of phase element of the impedance.

AC impedance has been widely used in the study of more complex systems, *e.g.* corrosion with and without passive layers. The equivalent circuits, however, grow in complexity and the interpretation becomes more difficult. Uncertainty about conclusions must increase as the number of fitting parameters expands. Moreover, other presentations of the data become preferable. These stories, however, fall outside the scope of this book although further details can be found in the Further Reading section at the end of the chapter.

7.4   MICROELECTRODES

In the discussion of the RDE and RRDE it is always assumed that the flux of reactant over the surface of the disc is totally uniform. In reality, this is not entirely true because there is an edge effect around the perimeter (Section 6.4.1). This is often visible when metal deposition is carried out at a potential where the deposition is mass transport controlled; the heavier deposit at the perimeter quickly becomes obvious. The theory for voltammetry at a RDE or RRDE is, however, appropriate because the current from the centre of the disc totally dominates and the contribution from the perimeter region is minor. If the disc electrode is shrunk so that its dimensions are $<100\,\mu m$, the opposite limit is reached; in steady state experiments, the contribution of the perimeter will dominate that from the centre. We have reached the world of the microelectrode.

Microelectrodes are generally defined as electrodes where at least one dimension is small enough that their properties are a function of size. Several geometries have been developed (*e.g.* discs, hemispheres, rings, lines) but microdiscs are the easiest to fabricate (*e.g.* by sealing microwires into glass or epoxy resin). Hence, the discussion here will be limited to microdiscs; their radii can vary between 0.1 and $100\,\mu m$ with perhaps 5–$25\,\mu m$ being the most popular range. Microelectrode experiments are almost always carried out in a still solution where diffusion is the only form of mass transport.

In a steady state experiment at a microdisc electrode, the dominant mass transport is through the $90°$ angle to the edge of the disc and the

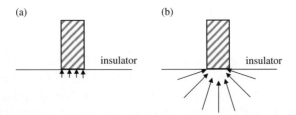

Figure 7.35 Diffusion to a microdisc electrode: (*a*) the planar diffusion field at short times and (*b*) the hemispherical diffusion field in the steady state.

microdisc is surrounded by a hemispherical diffusion field (Figure 7.35); the flux of reactant is proportional to the inverse of the microdisc radius. In short timescale experiments where the diffusion layer thickness remains thin compared to the dimensions of the microdisc, linear diffusion to a plane electrode is the mode of mass transport. Hence, the response of a microdisc electrode, radius, a, to a potential step from a value where $j = 0$ to one where the electrode reaction is diffusion controlled is given by:

$$j = \frac{nFD^{1/2}c}{\pi^{1/2}t^{1/2}} + \frac{4nFDc}{\pi a} \tag{7.73}$$

The first term is identical to that for a potential step under conditions of linear diffusion to a plane electrode and will dominate at short times. The second results from the hemispherical diffusion field and will dominate at long times and will certainly be the only term necessary in the description of the steady state. There will clearly be an intermediate time range where both terms are necessary for an accurate estimate of the current density. In most experiments, however, the timescale is chosen so that only one of the terms need be considered in the description of the experiment.

7.4.1 Steady State Experiments

After a relatively short time under any experimental conditions, a steady state, hemispherical diffusion field will be set up around the microdisc electrode (Figure 7.35). The mass transfer coefficient is given by:

$$k_{\mathrm{m}} = \frac{4D}{\pi a} \tag{7.74}$$

As a result, the limiting current density is given by:

$$j_{\mathrm{L}} = \frac{4nFDc}{\pi a} \tag{7.75}$$

and the limiting current by:

$$I_L = 4nFDca \tag{7.76}$$

The parameter determining the rate of diffusion is the size of the microdisc. By substituting values into Equations (7.8) and (7.74) it can readily be seen that the rate of mass transfer to a $2\,\mu m$ diameter microdisc is comparable to a RDE with a high rotation rate, $\sim 5000\,rpm$. Hence, steady state voltammetry at a microdisc electrode provides a powerful method for the study of rapid electron-transfer reactions. Voltammetry at a microdisc electrode, when carried out at relatively slow scan rates, is effectively a steady state method and therefore leads to sigmoidal shaped voltammograms similar to those recorded at a RDE. Indeed, the concepts can be directly transferred and the same equations may be used after allowing for a change in mass transfer coefficient from that of Equation (7.8) to that given by Equation (7.74).

The popularity of microelectrodes arises partly from the simple equipment required and partly because of the wide variety of media that can be used. The actual current passing through the cell with a microdisc is always low (often a few nA) and it is found that problems of *IR* distortion of voltammograms are much reduced. This greatly aids experiments where the current density is high or the conductivity of the electrolyte is low. Hence, mechanistic studies with high concentrations of reactant (as found in industrial electrolysis) become possible and it is also possible to study reactions in water without electrolyte, other low ionic strength solutions and rather resistive media.

7.4.2 Non-steady State Experiments

When short timescale experiments are carried out at a microelectrode, the mass transport regime is again linear diffusion to a plane electrode. Indeed, for this time regime, the conditions are identical to those at larger electrodes and the theoretical expressions discussed in Section 7.3 may be used without modification.

The use of microelectrodes, however, is highly advantageous for two practical reasons (Chapter 6): (a) *IR* distortion is much less of a problem and this allows the monitoring of much higher current densities without degradation of the responses and (b) interference from double layer charging currents is much reduced. These factors combine to make possible both experiments on a sub-millisecond timescale and cyclic voltammetry at high scan rates. Cyclic voltammetry with scan rates up to $1000\,V\,s^{-1}$ becomes relatively straightforward, with higher scan rates possible with care and experience.

7.5 SPECTROELECTROCHEMISTRY

Almost all investigations of electrode reactions would benefit from a spectroscopic record of the changes at the interface to supplement the information from voltammetry or other electrochemical experiments. In principle, it allows the positive identification of intermediates and/or the determination of the way their concentrations change with time or potential. As a result, it is not surprising that during the past 35 years there have been intense efforts to develop *in situ* spectroelectrochemical techniques. Experiments have, for example, been based on UV/visible, IR, Raman, ESR and, more recently, NMR spectroscopies and both X-ray absorption and surface X-ray scattering spectroscopies using synchrotron radiation. Also, the variants include both reflectance and transmission procedures with the beam either passing through the solution or from the back of a transparent electrode.

The task is always to discriminate the relative small changes at the electrode/solution interface from the spectroscopic response of the bulk electrolyte solution. Hence, the challenge is to design the experiment to increase the sensitivity of the technique to the electrode/electrolyte interface and also to discriminate between solution-free and surface bound intermediates. Several approaches have been used to increase the sensitivity to the electrode/electrolyte interface. These include (a) allowing the optical beam to pass only through a very thin layer of solution; (b) modulating the chemistry at the electrode with a known frequency (*e.g.* by using a potential pulse sequence to switch on and off the formation of an intermediate) and using a phase sensitive detector to identify the part of the total spectroscopic response that changes with the same frequency so as to record the difference in the spectra at the two potentials; and (c) repeating an experiment many times and signal averaging the spectroscopic response. Unfortunately, almost always the spectroelectrochemical experiment requires a compromise in the cell design that, to some extent, degrades the electrochemical response (*e.g.* pushing the electrode up against an optical window must produce a non-uniform current distribution). In addition, the experiments usually require special equipment. Hence, no spectroelectrochemical experiment has found general utility; while most of the techniques have had their successes, their application is usually limited to a few laboratories that have been prepared to devote time and effort to their development. In this book, spectroelectrochemistry will be illustrated with a couple of examples.

An early spectroelectrochemical experiment requiring only relatively simple equipment employs an optically transparent thin layer cell (an

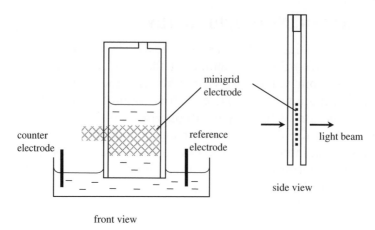

Figure 7.36 Schematic of an electrochemical thin layer cell incorporating an optically transparent electrode (OTTLE).

OTTLE) (Figure 7.36). The electrode is fabricated from a metal (commonly gold) minigrid, a mesh made from very thin wires and having 100–2000 perforations per inch and capable of transmitting >80% of visible light. This electrode is sandwiched between two microscope slides held together by a ~100 μm thick adhesive spacer (or separated by an inert polymer spacer and sealed with epoxy resin). The OTTLE is placed in a beaker that also contains the reference and counter electrode and fills by capillary action. The working electrode is placed directly in the beam of a visible spectrometer. Clearly, the cell has a poor electrode geometry, see Chapter 6, and it is therefore not suitable for short timescale experiments. Rather it is suited to steady state experiments where the spectra are recorded after a potential has been applied for a period long enough for all of the solution in the OTTLE to come to an equilibrium; The species monitored also needs to have a strong chromophore. OTTLEs have been widely used to determine n values and to determine formal potentials for biologically active compounds. In such experiments, a series of potentials is applied to the minigrid electrode. At each potential, the charge is measured for the current to decay to the background level, usually taking <60 s, when the electrolyte will contain the equilibrium ratio of c_O/c_R for the applied potential and the potential measured *versus* the reference electrode will be free of significant IR drop (since the current is close to zero). A spectrum is recorded and used to calculate the ratio of c_O/c_R. The measured potential and the calculated value of c_O/c_R are then used to make a plot of the Nernst equation and the formal potential is

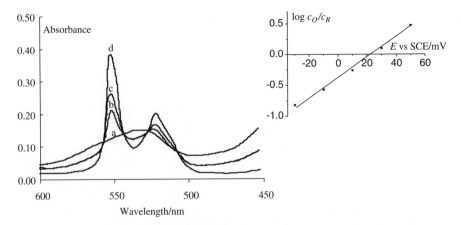

Figure 7.37 Spectra using a thin layer cell containing a solution of 0.4 mM cyto-chrome c + 2,6-dichlorophenolindophenol + 0.1 M phosphate buffer, pH 7 after electrolysis at (a) + 250, (b) + 30, (c) + 10 and (d) –250 mV *vs* SCE. Inset: plot of the Nernst equation to determine the formal potential of the cytochrome couple, + 22 mV *vs* SCE. Data taken from W. R. Heineman, B. J. Norris and J. F. Goelz, *Anal. Chem.*, 1975, **47**, 79.

determined by extrapolation to $c_O/c_R = 1.0$. The value of n is conveniently obtained from the slope of the Nernst plot or by plotting c_O or c_R *vs* the total charge passed. Figure 7.37 shows data for the determination of the formal potential for cytochrome c using 2,6-dichlorophenolindophenol as a mediator. Clearly, as the cytochrome c is reduced, sharp peaks grow at 520 and 550 nm. When the Nernst plot is drawn, it can be seen that the slope is 59 mV (the cytochrome c undergoes a reversible $1e^-$ reduction) and the couple has a formal potential of + 21 mV *vs* SCE. Changing the potential in a positive direction, the data could be reproduced, showing that both the oxidized and reduced form of cytochrome c are stable and that the couple is fully reversible.

A more sophisticated spectroelectrochemical experiment is potential modulated, external reflectance infrared spectroscopy (EMIRS). The experiment requires purpose-built instrumentation but is capable of giving spectra with molecular specificity for adsorbed intermediates. Interference from the electrolyte is minimized by (a) pushing the electrode against an optical window so that the infrared beam passes through only a thin layer ($\sim 10\,\mu m$) of trapped electrolyte solution and (b) modulating the potential between two values and recording the difference spectrum (the spectroscopic response of the bulk solution will be identical at both potentials). The change in spectrum between potentials is small and hence the modulation is continuous and signal averaging is employed to improve the response. The application of the

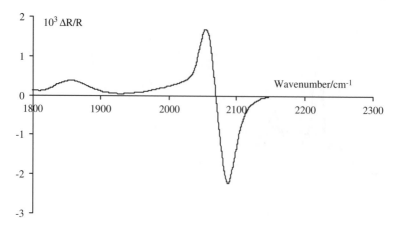

Figure 7.38 EMIR spectrum for the adsorbate on a Pt electrode in 0.5 M CH$_3$OH + 1 M H$_2$SO$_4$. Modulation between + 50 and + 450 mV *vs* SHE with a frequency of 8.5 s^{-1}. Data taken from B. Beden, C. Lamy, A. Bewick and K. Kinimatsu, *J. Electroanal. Chem.*, 1981, **121**, 343.

technique is illustrated using the example of the chemical adsorption of methanol on platinum from 1 M H$_2$SO$_4$ + 0.5 M CH$_3$OH. Figure 7.38 shows an EMIR spectrum when the potential is modulated between + 50 and + 450 mV *vs* RHE at 8.5 s^{-1} (averaged over 100 scans). The spectrum shows responses at 2060 and 1840 cm^{-1}, typical for linear bonded CO and bridge bonded CO species, respectively. The spectral response at 2060 cm^{-1} has the 'differential peak' shape because the linear bonded CO is adsorbed at both potentials but the strength of the bond and hence the absorption maximum shifts slightly with potential. The form of the response at 1840 cm^{-1} indicates that the bridge bonded CO exists predominately only at 0 mV. If the modulation of potential is changed to between 0 and + 700 mV, the IR spectrum disappears as the CO intermediate is oxidized at + 700 mV. It may be concluded that adsorbed CO is the surface poison that prevents facile oxidation of methanol to carbon dioxide at low positive potentials.

Further Reading

1. Southampton Electrochemistry Group, *Instrumental Methods in Electrochemistry*, Ellis Horwood, Chichester, republished 2001.
2. A. J. Bard and L. R. Faulkner, *Electrochemical Methods*, John Wiley & Sons, New York, 2001.
3. *Laboratory Techniques in Electroanalytical Chemistry*, ed. P. T. Kissinger and W. R. Heineman, Marcel Dekker, New York, 1996.

4. Z. Galus, *Fundamentals of Electrochemical Analysis*, Ellis Horwood, Chichester, 1994.

5. *Electrode Kinetics, Principles and Methodology*, ed. C. H. Bamford and R. G. Compton, Comprehensive Chemical Kinetics, Vol. 26, Elsevier, Amsterdam, 1986.

6. O. Hammerich, in *Organic Electrochemistry*, ed. H. Lund and O. Hammerich, Marcel Dekker, New York, 2001, p. 95.

7. D. D. MacDonald, *Transient Techniques in Electrochemistry*, Plenum, New York, 1977.

8. *Impedance Spectroscopy – Theory, Experiment and Application*, ed. E. E. Barsoukov and J. R. Macdonald, John Wiley & Sons, New York, 2005.

9. M. Orazem, Notes for ECS Short Course, *Advanced Impedance Spectroscopy*, http://orazem.che.ufl.edu/Downloads.html, Jan 2009.

10. M. E. Orazem and B. Tribollet, *Electrochemical Impedance Spectroscopy*, ECS – Wiley, New York, 2008.

11. *Ultramicroelectrodes*, ed. M. Fleischmann, S. Pons, D. R. Rolinson and P. Schmidt, Datatech Science, Morganton, NC, 1987.

12. *Microelectrodes: Theory and Applications*, ed. M. I. Montenegro, M. A. Queiros and J. L. Daschbach, NATO ASI Series E197, Kluwer Academic, Dordrecht, 1991.

13. *New Techniques for the Study of Electrodes and their Reactions*, ed. R. G. Compton and A. Hamnett, Comprehensive Chemical Kinetics, Vol. 29, Elsevier, Amsterdam, 1989.

14. *Techniques for Characterisation of Electrodes and Electrochemical Processes*, ed. R. Varma and J. R. Selman, The Electrochemical Society, New York, 1991.

15. *Spectroelectrochemistry – Theory and Practice*, ed. R. J. Gale, Plenum, New York, 1988.

16. *Electrochemical Interfaces*, ed. H. D. Abruna, VCH, Weinheim, 1991.

17. R. G. Compton and C. E. Banks, *Understanding Voltammetry*, World Publishing Co., Singapore, 2007.

CHAPTER 8
Fuel Cells

8.1 INTRODUCTION

Our society faces many challenges if our present lifestyle is to continue and develop and also spread to the population throughout the world. There can be no doubt that our reserves of fossil fuel are finite and, at least, long term we need to harness renewable energy sources such as solar, wind and tidal energy. More immediately, we can benefit from using fossil fuels more efficiently and finding ways to utilize alternative sources of carbon from plants. Another pressing requirement is to find ways to reduce our emissions of carbon dioxide and again this can be achieved only by using fossil fuels more efficiently, promoting the use of renewable energy sources and meeting our desire to move around using technology that does not lead to emissions of carbon dioxide.

Electrochemistry and, more particularly fuel cells, can contribute to solving these problems in several diverse ways. Fuel cells can be a primary source of energy and could provide an efficient way to generate energy from fuels, either on a large scale or locally at the point of use at a single customer level. While electricity is a convenient way to distribute energy, a national energy system based on renewable sources of energy must have a storage buffer between the generation and the consumer. Nature does not provide sunlight, wind or tidal power at the whim of the consumer and the electricity produced must be stored and made available when the customer demands it. The combination of water electrolysis, H_2 storage and a fuel cell is one possible solution to this

A First Course in Electrode Processes, 2nd Edition
By Derek Pletcher
© Derek Pletcher 2009
Published by the Royal Society of Chemistry, www.rsc.org

problem. Mobile sources of energy for vehicles, trains, planes and ships are also essential and again fuel cells could meet this need.

8.2 WHAT IS A FUEL CELL?

A fuel cell is a device that converts chemical energy into useful electrical energy; the overall cell reaction must be spontaneous (*i.e.* have a negative Gibbs free energy) and will involve a redox reaction occurring *via* an oxidation and a reduction at the anode and cathode, respectively. In contrast to a driven electrolysis cell, the cathode will be the positive electrode and the anode the negative electrode. A fuel cell is distinguished from a battery by the fact that the reactants for the two electrode reactions are stored externally to the cell. In consequence, the energy storage capability is not determined by the cell performance but is defined by the size of the external storage tanks and the ease with which these may be refilled/replaced.

A fuel cell or a collection of fuel cells (a stack or, on an even larger scale, a system consisting of many stacks) is always designed for a specific purpose, *i.e.* to provide power for a PC, a mobile phone, a vehicle, a hotel, a city, *etc.* Therefore, the power output required may vary from mW through kW to MW. Power is defined by the equation:

$$\text{Power} = IV_{\text{cell}} \tag{8.1}$$

and has the units of watts (W). Depending on the application, power is also quoted in W per unit weight or W per unit volume, and in the laboratory it is often discussed in terms of W per unit area of electrode (apparent geometric area of the electrodes).

The performance of a fuel cell system is therefore determined by the cell voltage and current density and these will be stressed throughout this chapter. In addition to influencing the power output, it should be recognized that the cell voltage reflects the effectiveness of the fuel cell system in extracting the chemical energy stored in the fuels and converting it into electrical energy. The cell voltage will be given by:

$$V_{\text{cell}} = \Delta E_{\text{e}} - \eta_{\text{a}} - \eta_{\text{c}} - IR \tag{8.2}$$

where ΔE_{e} is the difference in the equilibrium potentials of the anode and cathode reactions and it is related to the Gibbs free energy of the overall cell reaction by:

$$\Delta G = -nF\Delta E_{\text{e}} \tag{8.3}$$

and η_a and η_c are the overpotentials associated with the two electrode reactions while the final term is the *IR* drop through the cell, *i.e.* through the electrolyte between the electrodes, the electrodes, contacts, *etc.* The overpotentials and *IR* drops must be regarded as inefficiencies lowering the fraction of chemical energy converted into electrical energy and resulting in heat. Hence, the energy efficiency, the % of the theoretical energy that is actually obtained, is defined by:

$$\text{Energy efficiency} = \frac{V_{cell}}{\Delta E_e} \cdot 100 \qquad (8.4)$$

The economic success of the fuel cell system will also be determined by the cost of the components, especially the electrocatalysts (usually Pt metals), the electrolyte, the gas feed distributors and the bipolar plates that interconnect neighbouring cells in the stack. This will depend on both the materials employed and the design of the components being amenable to simple mass fabrication procedures. In addition, the lifetime of components is an issue and maintaining a high current density and a high cell voltage over a period of several years is essential. Depending on the application, compactness and/or weight may also be important.

In almost every type of fuel cell, the cathode reaction is the reduction of oxygen:

$$O_2 + 4H^+ + 4e^- \longrightarrow 2H_2O \qquad (8.5)$$

although it is clearly advantageous that the cell feed is the surrounding air. It is important that the full $4e^-$ reduction takes place. Any formation of hydrogen peroxide will decrease the magnitude of the negative Gibbs free energy and therefore V_{cell}; in consequence, both the power output and the efficiency of conversion of fuel into electrical energy will be decreased.

Several fuels are considered. The most common is hydrogen:

$$H_2 - 2e^- \longrightarrow 2H^+ \qquad (8.6)$$

The great attraction of hydrogen is that the only product from a H_2/O_2 fuel cell is water (and no greenhouse gas). The drawback is that hydrogen is not a primary fuel and must be made by another chemical process, either at a large facility and then distributed or locally at the fuel cell. For example, some 'methanol fuel cells' include a reformer to convert methanol into hydrogen that is then fed to the fuel cell. Alternatively, the fuel cell anode reaction could be the direct oxidation of methanol:

$$CH_3OH + H_2O - 6e^- \longrightarrow CO_2 + 6H^+ \qquad (8.7)$$

A consequence of using methanol or any other organic fuel is, however, that a product is carbon dioxide; a fuel cell is then only an advance on other technologies (*e.g.* burning) if the fuel cell provides a better a efficiency for the conversion of the chemical energy into electrical energy and/or less carbon dioxide is emitted per watt of power produced. Ethanol (possibly derived from biosources) or hydrocarbons:

$$C_2H_5OH + 3H_2O - 12e^- \longrightarrow 2CO_2 + 12H^+ \qquad (8.8)$$

$$CH_4 + 2H_2O - 8e^- \longrightarrow CO_2 + 8H^+ \qquad (8.9)$$

are other attractive fuels provided that the fuel cell competes in efficiency with other technologies. Table 8.1 summarizes the overall cell reactions with each of these fuels and their equilibrium (maximum) cell voltages estimated from thermodynamics. The similarity in the equilibrium cell voltages is coincidence but noteworthy.

A much larger influence on the maximum power output is the maximum current density that can be achieved; the target will be a current density of $>1\,A\,cm^{-2}$. To approach this target two problems must be overcome. Firstly, it will be necessary to have a high rate of mass transport of the reactants to the site of electron transfer. With gaseous reactants, *e.g.* O_2 and H_2, this is impossible when they are dissolved in the electrolyte because of their low solubility in aqueous media. Obtaining a sufficient flux of reactants has required the development of gas diffusion electrodes (GDEs) where the gas is fed from the back of the electrode through a porous structure to the electrode/solution interface; such electrodes will be discussed further later in this chapter. Secondly, electrocatalysts are essential to obtain a high rate of the electrode reactions at low overpotentials (Chapter 5). In general, the catalysts are fabricated into a highly dispersed state on an inert substrate such as a carbon powder and this forms part of the structure of the GDE. The highly dispersed state is partly to make optimum use of expensive catalysts, which are precious metals in many fuel cells, but also to

Table 8.1 Overall cell reactions for four possible fuel cell chemistries and their equilibrium (maximum) cell voltages estimated from thermodynamics at ambient temperature.

Fuel	Overall fuel cell reaction	$\Delta E_e/mV$
H_2	$2H_2 + O_2 \longrightarrow 2H_2O$	1223
CH_3OH	$CH_3OH + 1\frac{1}{2}O_2 \longrightarrow CO_2 + 2H_2O$	1220
C_2H_5OH	$C_2H_5OH + 3O_2 \longrightarrow 2CO_2 + 3H_2O$	1140
CH_4	$CH_4 + 2O_2 \longrightarrow CO_2 + 2H_2O$	1061

present a high surface area of electrocatalyst to the solution. As noted in Chapter 5, defining good electrocatalysts for the reduction of oxygen is difficult and overpotentials on the best catalysts can often approach 500 mV, representing a substantial loss in energy efficiency when the equilibrium cell potential is only ~ 1200 mV.

Using an elevated temperature is helpful in reducing overpotentials. In common with other chemical reactions, the rate of kinetically controlled electrode reactions approximately doubles every 10 K. If the resulting materials problems can be solved, this is a good reason to operate the fuel cell at a higher temperature. In addition, elevating the temperature will also increase the conductivity of the electrolyte, reducing *IR* losses.

8.3 TYPES OF FUEL CELL

Several types of fuel cells continue to undergo development and testing. They differ in the type of electrolyte and their operating temperatures. It will be seen that, ideally, the electrolyte should transport a single ion. Certainly, it must be a poor electronic conductor since electron conduction would lead to shorting of the two electrodes.

8.3.1 Phosphoric Acid Fuel Cells

The phosphoric acid fuel cell was one of the first fuel cells, dating back to the early 1970s. It is also the most developed of the technologies. The electrode reactions are (8.5) and (8.6) and the basic cell is shown in Figure 8.1. The electrolyte is 98% phosphoric acid absorbed into a porous, solid ceramic matrix made by heating 1–5 μm SiC powder with PTFE binder. The use of this very concentrated acid allows operation at temperatures up to 470 K, improving the kinetics of O_2 reduction and decreasing the problem of CO poisoning of the anode catalyst (due to carbon containing impurities in the H_2 feed) as well as improving electrolyte conductivity. The catalysts are usually Pt in a high area form dispersed onto a carbon powder support.

The two gas diffusion electrodes are multilayer structures (Figure 8.2a). A carbon paper or carbon cloth, treated to become hydrophobic, provides structural support and acts as a local current collector. On the support is a carbon powder/PTFE layer designed to be hydrophobic and highly porous to gas. On top of this and exposed to the electrolyte is a thinner, less hydrophobic layer containing the dispersed Pt catalyst on

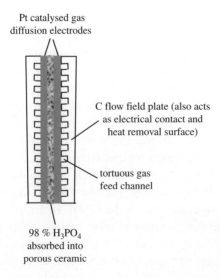

Pt catalysed gas
diffusion electrodes

C flow field plate (also acts
as electrical contact and
heat removal surface)

tortuous gas
feed channel

98 % H_3PO_4
absorbed into
porous ceramic

Figure 8.1 Schematic representation of a phosphoric acid fuel cell.

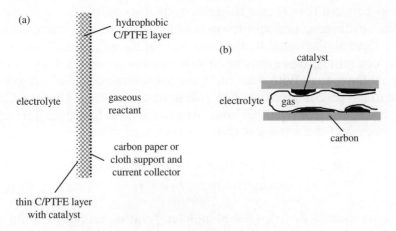

(a)

hydrophobic
C/PTFE layer

(b)

catalyst

electrolyte

gaseous
reactant

electrolyte   gas

carbon

carbon paper or
cloth support and
current collector

thin C/PTFE layer
with catalyst

Figure 8.2 (*a*) Typical structure of a gas diffusion electrode and (*b*) a simple model of
a pore within the catalyst layer.

carbon powder. The properties of these layers are controlled through the
PTFE content (controlling the hydrophobicity) and the size, porosity,
wettability and other properties of the carbon powder employed. The
electrode is designed to give a high area of three-phase interface between
the phosphoric acid, catalyst and gaseous reactant. Hence, the correct
level of penetration by both electrolyte and gases is necessary but neither
must pass through the structure. The PTFE is present to make the
structure hydrophobic and to resist deep penetration by electrolyte.
Figure 8.2(b) shows a simple model for the three-phase interface within

the porous layer. This model is based on the concept of an ultra-thin layer of electrolyte over the catalyst centres so that there is very rapid transport of gas through this layer and also rapid exchange between the gas and electrolyte because of the high interfacial area. Clearly, the catalyst will be active only where the appropriate three-phase interface exists and hence the effective usage of the expensive catalysts requires that catalyst is restricted to where the interface can be formed. The catalysts are precious metals because these are among the few materials stable to corrosion in the acidic medium at elevated temperature.

Behind each electrode is a carbon block (usually termed the flow field plate) with channels in the face on the GDE side to give a uniform supply of the gaseous reactants to the GDE. In addition, this component serves as the current contact and also to remove water vapour and excess heat from the system. In a stack, the cell of Figure 8.1 will be repeated many times and these are usually connected in series to give a voltage output (maybe 120 or 240 V). The carbon plate then also serves as the electrical connection between neighbouring cells as well as acting as a barrier between the O_2 and H_2 gases in the two cells.

Pure hydrogen, perhaps from a renewable energy source driving water electrolysis, would be the ideal fuel for phosphoric acid fuel cells. Most commonly, however, the primary fuel is natural gas or another carbonaceous fuel. With such fuels, the phosphoric acid fuel cell system must then be combined with a 'catalytic converter'. This is a unit external to the fuel cells that converts the fuel into hydrogen *via* steam reforming and the water-gas shift reaction, *e.g.*:

$$CH_4 + H_2O \longrightarrow CO + 3H_2 \tag{8.10}$$

$$CO + H_2O \longrightarrow CO_2 + H_2 \tag{8.11}$$

These reactions are carried out at high temperature using palladium and cobalt catalysts, respectively, presently in sequential reactors although there is development of co-catalysts that combine the two steps. Under the conditions of the phosphoric acid fuel cell, the anode catalyst can tolerate up to 1.5% CO in the H_2 feed; if this level is exceeded, it is necessary to remove CO. Any sulfur gases from impurities in the primary fuel must also be removed.

Typically, the cell voltage is $\sim 0.65\,\text{V}$ at a current density of $0.25\,\text{A}\,\text{cm}^{-2}$, giving a maximum power output of $\sim 0.15\,\text{W}\,\text{cm}^{-2}$. Phosphoric acid fuel cells were targeted at stationary applications, either large scale electricity generation for a utility company or smaller units to provide power to a hospital, hotel, school, *etc.* More than a hundred phosphoric acid fuel cell systems have been installed in 19 countries.

Several hundred 100–200 kW units have been installed but the largest is an 11 MW plant in Japan; it consists of 18 stacks each containing 492 cells with $0.93\,m^2$ electrodes. The overall energy efficiency (natural gas to electrical energy) is $\sim 40\%$ compared with an efficiency of $\sim 35\%$ for a modern thermal power generation plant (in both cases, the efficiency almost doubles if the heat generated is used in a cogeneration plant). In view of the cost of the fuel cells (2–3 times higher than conventional generation technology) and the modest improvement in energy efficiency compared to conventional plant, phosphoric acid fuels cells become commercially competitive only in special circumstances such as a remote location with a limited requirement for electricity.

8.3.2 Alkaline Fuel Cells

Alkaline fuel cells again employ hydrogen as the fuel and the cell structure is similar to that of the phosphoric acid fuel cell. The GDEs are also based on similar concepts although the carbon paper or cloth support might be replaced by a Ni gauze (stable to corrosion in the alkaline environment). The electrolyte is 30% KOH at a temperature approaching 373 K. In some cell designs, the electrolyte is again absorbed into a porous ceramic material. In others, the electrolyte is flowed through the interelectrode gap. This approach was developed to allow the use of a larger volume of electrolyte and hence to decrease the rate at which the electrolyte becomes exhausted by the conversion of hydroxide into carbonate by absorption of carbon dioxide either from air fed to the cathode or as impurity in the hydrogen fuel. Even so, it is advantageous to remove the CO_2 from the air supply to the cathode, and using the product from a catalytic converter as feed to the anode is certainly ruled out (CO_2 is the major C product from the natural gas or methanol feedstock, see above). Pure H_2 must be used.

Because of the high pH, the electrode reactions are correctly written:

$$O_2 + 2H_2O + 4e^- \longrightarrow 4OH^- \tag{8.12}$$

$$H_2 + 2OH^- - 2e^- \longrightarrow 2H_2O \tag{8.13}$$

The attractive feature of this technology is the possibility of operating without platinum metal catalysts in the alkaline environment, thereby greatly reducing costs. Electrocatalysts based upon both Raney nickel and silver catalysed carbon materials have been successfully tested although some developers still opt for Pd or Pt catalysts. In addition, alkaline fuel cells have a much higher energy efficiency, $\sim 70\%$, than the acid systems, largely because of the better kinetics of O_2 reduction in strongly alkaline media. A current density of $400\,mA\,cm^{-2}$ can be achieved with a cell voltage of $\sim 0.8\,V$, reaching a power output of $0.32\,W\,cm^{-2}$. Alkaline fuel cells were developed as early as the 1960s and were utilized in the Apollo space programme and units with a power output of 1–$100\,kW$ have found niche application. The difficulty of electrolyte exhaustion due to CO_2 absorption has limited further development but interest is rekindled periodically because of the outstanding energy efficiency and the possibility of low cost catalysts.

8.3.3 Polymer Electrolyte Membrane (PEM) Fuel Cells

In these fuel cells, a proton conducting polymer replaces the aqueous electrolyte. Usually, the membrane and two electrocatalyst layers are fabricated into a single component known as the membrane electrode assembly or MEA. The preferred fuel is pure hydrogen. Methanol can also be used as a fuel either directly at the anode or indirectly by conversion into hydrogen in an external catalytic reformer. If used directly in the fuel cell, it is necessary to use an anode with a higher catalyst loading and to operate at lower current density. The present generation of H_2/O_2 PEM fuel cells are operated at a temperature of 333–$353\,K$ although a goal of many R & D programmes is to increase the operating temperature to $\sim 410\,K$ to enhance the kinetics of O_2 reduction and decrease the poisoning of the catalysts by impurities in the gas feeds. This technology is discussed further in Section 8.4.

8.3.4 Molten Carbonate Fuel Cells

In the molten carbonate fuel cell the electrolyte is a molten salt, a carbonate eutectic (either K_2CO_3/Li_2CO_3 or Na_2CO_3/Li_2CO_3), which requires a very elevated temperature of $\sim 900\,K$. Such a high temperature leads to a requirement for quite different materials for the flow field plate, current collectors and gaskets as well as electrode materials and such materials are still under development. Since the materials are so different from those of low-temperature fuel cells, the design and engineering of molten carbonate fuel cells are also different. Both parallel plate and tubular designs have been operated; tubular designs are often more compatible with the use of ceramic materials. On the other hand, the high temperature enhances substantially the kinetics of all reactions, introducing the possibility of new chemistry and cheaper catalysts. The electrolyte is absorbed in a thin porous ceramic tile (0.5–1 mm thick), commonly made from γ-LiAlO$_2$, and the electrode materials are coated onto the two faces of the tile. The cathode is lithiated NiO and the anode a NiCr alloy and these layers are deposited onto opposite faces of a ceramic tile or tube. The cathode reaction is:

$$O_2 + 2CO_2 + 4e^- \longrightarrow 2CO_3^{2-} \tag{8.14}$$

while, because of the high temperature, the anode reaction can be the oxidation of H$_2$, CO, natural gas or other carbonaceous fuels, *e.g.*:

$$H_2 + CO_3^{2-} - 2e^- \longrightarrow CO_2 + H_2O \tag{8.15}$$

$$CO + CO_3^{2-} - 2e^- \longrightarrow 2CO_2 \tag{8.16}$$

$$CH_4 + 4CO_3^{2-} - 8e^- \longrightarrow 5CO_2 + 2H_2O \tag{8.17}$$

Some view the anode reactions as occurring by reforming of all the fuels to hydrogen within the electrocatalyst layer. In some versions of the technology, a separate layer of reforming catalyst within the cell structure is employed. It can be seen that the operation of the cell requires the transport of carbonate through the electrolyte and also the transfer of carbon dioxide from the anode to the cathode though ducts external to the cell. While the technology is complex, it has the advantage that no Pt metal catalysts are employed and CO and hydrocarbons are possible feedstocks. The thermodynamic cell potential clearly depends on the fuel but typically a molten carbonate fuel cell can achieve a cell voltage of 0.88 V at a current density of 150 mA cm^{-2} and the energy efficiency is $\sim 50\%$.

Molten carbonate fuel cells are targeted towards larger stationary power supplies by utility companies because the system design is too complex for small installations. Several prototype units with power outputs of 5–25 kW have been produced and a 1 MW and a 2 MW unit have been operated in Japan and the USA, respectively. The technology still attracts development because of the flexibility in the choice of fuel, the absence of precious metal catalysts and the potential to develop improved performance materials.

8.3.5 Solid Oxide Fuel Cells

This is the youngest of the technologies and it is undergoing intensive development. Presently, solid oxide fuel cells operate at even higher temperatures, typically 1220 K but, with the development of suitable materials, the operating temperature might be reduced to perhaps 900 K. The electrode reactions are again different. At the cathode:

$$O_2 + 4e^- \longrightarrow 2O^{2-} \tag{8.18}$$

and at the anode (depending on the fuel):

$$H_2 + O^{2-} - 2e^- \longrightarrow H_2O \qquad (8.19)$$

or:

$$CH_4 + 4O^{2-} - 8e^- \longrightarrow CO_2 + 2H_2O \qquad (8.20)$$

Again, because of the high temperature, natural gas, other carbonaceous fuels and CO can be employed directly as fuels although the mechanism may involve reforming within the electrocatalyst layer. The electrolyte must conduct oxide ions from cathode to anode. Therefore, the cell consists of a tile or tube (1–2 mm thick) of an oxide conducting solid oxide (*e.g.* $Zr_{0.9}Y_{0.1}O_{1.95}$ or $Ce_{0.9}Gd_{0.1}O_{1.95}$) with electrocatalyst layers on each side. The anode catalyst is usually a Ni-ZrO_2 cermet and the cathode catalyst a mixed oxide (*e.g.* $La_{0.9}Sr_{0.1}MnO_3$). Solid oxide fuel cells are much more tolerant to impurities than low temperature fuel cells. Again this technology is complex, but it also has the advantage that no Pt metal catalysts are employed and a diversity of fuels can be used.

   While maintaining a cell voltage of ~ 0.7 V, a solid oxide fuel cell would give a current density of 0.2 A cm^{-2}, fed with natural gas, or 0.4 A cm^{-2} fed with H_2. The oxide-conducting electrolyte has a substantial resistance and the development of materials with a higher conductivity or better mechanical properties to allow the use of thinner membranes is a priority. Solid oxide fuel cells are being developed for stationary power supplies and auxiliary power units for trains and ships. Demonstration units with a power output of 250 kW have been built but commercial systems must await improvements in materials and cell design.

8.4 H_2/O_2 PEM FUEL CELLS

H_2/O_2 PEM fuel cells are the most promising of the technologies for the power units for light-duty electric vehicles (*e.g.* cars) because of their compactness, absence of hazardous liquid electrolytes, relatively low operating temperature and output characteristics (*i.e.* high power and capability to change power output rapidly). In addition, they could be used for remote power generation and for replacement for small batteries in, for example, computers. A sketch of a PEM fuel cell would look similar to Figure 8.1 but, in reality, the electrocatalyst layers are fabricated onto the polymer electrolyte layer to give a single component termed the MEA (membrane electrode assembly). Initially, the discussion will assume that the fuel is pure hydrogen.

8.4.1 Solid Polymer Electrolyte (Membrane)

The roles of the membrane are:

- To allow the transport (migration) of protons from the anode to the cathode with the minimum voltage drop; the *IR* drop reduces the fuel cell voltage, see Equation (8.2), and hence both the energy efficiency and the power output.
- To maintain the separation of the hydrogen and oxygen gases (mixing would allow rapid, even explosive, reaction in the presence of the catalyst materials).
- To act as a support for the catalyst layers and allow the formation of an effective three-phase interface between the membrane polymer, electrocatalyst and reacting gas.

It must also be stable in the PEM fuel cell environment, *i.e.* strongly acidic with exposure to both oxidizing and reducing conditions, and have the mechanical and physical properties to be handled and used as a thin film.

The resistance of the membrane is proportional to its thickness and inversely proportional to the polymer conductivity. Hence, the goal is very thin membranes made from a polymer with a high proton conductivity. The proton conductivity results from anionic functional groups covalently bonded within the polymer (the fixed ionic groups); in the proton form of the polymers, the counter ions for charge balance are protons and transport of the protons occurs by protons hopping between fixed ionic groups under the influence of a voltage gradient. Clearly, the conductivity is strongly influenced by the chemical properties of the fixed ionic group and their 'concentration', quoted either as the equivalent weight of the polymer or its ion-exchange capacity. Hydration of the polymers is essential for high conductivity as it weakens the ionic interactions between the protons and fixed ionic groups. Indeed, the successful polymers have a high water content, as indicated by the substantial swelling (20–70%) when the dry membrane is contacted by water. Several factors determine the level of hydration:

- the chemical properties of the polymer backbone, particularly the degree to which it is hydrophobic/hydrophilic;
- the properties of the fixed ionic groups;
- the presence of crosslinking in the polymer structure; crosslinking prevents the swelling of the polymer and limits water uptake. The crosslinking may be covalent bonding between polymer chains (*e.g.*

in styrene/divinylbenzene copolymers, the crosslinking results from the presence of the divinylbenzene in the polymer structure) or interchain interactions (*e.g.* in the perfluorinated polymers, the perfluorinated backbones are thought to interact strongly so that the polymer undergoes phase separation into hydrophobic zones containing the fluorocarbon chains and hydrophilic zones where the fixed ionic groups and counter ions have collected together);

• the 'concentration' of the fixed ionic groups. This also fixes the 'concentration' of protons in the polymer. The 'concentration' of both anions and cations determines the hydration level of the polymer;

• the media on the two sides of the membrane; in the H_2/O_2 fuel cell, the critical factor is the vapour pressure of water in the two gases.

The extent of hydration possible is, however, limited by the tendency to 'dissolve' as a polyelectrolyte and, hence, to lose stability as a solid material, particularly on the longer timescale. Therefore, some compromises in the membrane structure must be accepted.

Most current PEM fuel cells are based on perfluorinated polymers with sulfonate fixed ionic groups (Figure 8.3). The most common is Nafion™, manufactured by DuPont. It is routinely marketed as membranes where the polymer has an equivalent weight of 1100 and a conductivity of ~ 0.12 S cm^{-1} at 340 K when the counter ions are protons. Membranes are supplied in several thicknesses, *e.g.* Nafion 117 and Nafion 112 are 180 and 50 μm thick, respectively. With a typical PEM fuel cell current of 0.5 A cm^{-2}, the *IR* drop through a Nafion 112

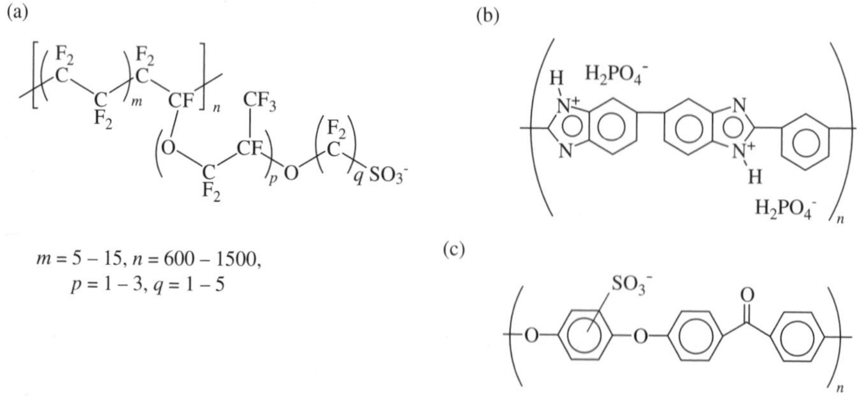

(a)   (b)

$m = 5 - 15, n = 600 - 1500,$
$p = 1 - 3, q = 1 - 5$

(c)

Figure 8.3   Polymers used in the development of PEM fuel cells: (*a*) perfluorinated polymer with side chains ending in sulfonate groups, (*b*) polybenzimidazole/phosphoric acid (PBI) and (*c*) sulfonated polyether ketone (PEEK).

membrane will be only $\sim 20\,mV$ under the above conditions. There are also membranes fabricated from lower equivalent weight polymer (therefore, having a higher conductivity) but they generally need to be slightly thicker to maintain long-term stability and, hence, do not lead to significant gain in fuel cell performance. These perfluorinated polymers with sulfonate fixed ionic groups have excellent performance for fuel cells operating at 330–350 K. The fluorinated polymers are, however, expensive and also show a decline in performance at higher temperatures because of an inability to maintain a high water content. There has, therefore, been an intensive effort to develop alternative membranes that are both cheaper and are suitable for the next generation of PEM fuel cells operating at up to 410 K. The cost can be reduced substantially only by employing polymers that are not fluorinated while, in general, polymers based on aromatic structures have a greater stability under fuel cell conditions. Two such polymers used in the development of PEM fuel cells are also shown in Figure 8.3. The first is poly(benzimidazole) protonated with phosphoric acid (PBI) and the other is a sulfonated poly(ether ketone) (PEEK). Both are claimed to give voltage drops similar to Nafion™ at $\sim 340\,K$ and to have the required lifetime. Moreover, both may be operated well above 373 K and their water retention can be improved further by fabricating polymer/inorganic composites. The most common inorganic additive is silica powder.

8.4.2 Cathode Catalyst

Defining improved electrocatalysts for oxygen reduction remains one of the great challenges. The mechanisms for oxygen reduction are discussed in Chapter 5. Here, it should be noted that the catalyst should support the full $4e^-$ reduction of oxygen, Equation (8.5), at low overpotentials (*i.e.* close to the equilibrium potential, $+1.23$ V *vs* SHE). In reality, even the best electrocatalysts now available operate with an overpotential of $>400\,mV$ and this is by far the biggest inefficiency in the fuel cell. Moreover, there are very few materials stable to corrosion or dissolution at the potential of the oxygen cathode ($\sim +0.8$ V *vs* SHE) in the acidic environment. As a result, it is almost impossible to avoid the use of a precious metal as the cathode catalyst and a major thrust of R & D is aimed towards reducing the amount of catalyst in the cathode. As the fuel cell cathode operates under conditions where the reduction of oxygen is kinetically controlled, the current density is proportional to the real area of the catalyst/electrolyte interface and it is therefore important that the catalyst is presented in a form where the surface area/gram of catalyst is as high as possible.

The industry standard for the electrocatalyst is platinum although the exchange current density for oxygen reduction at smooth Pt is only some $10^{-9}\,A\,cm^{-2}$. In the fuel cell, the performance is enhanced by creating very high surface area forms of Pt, but even then the overpotential is $>400\,mV$ at operating current densities. No other precious metals achieve even this overpotential. Alloying platinum with metals such as Cr, Fe and Mn reduces the overpotential by maybe $25\,mV$ but both Fe and Mn leach out in the acid environment and the resulting ions interact more strongly than protons with the sulfonate groups in the membrane, thereby increasing the membrane *IR* drop. There is interest in chromium alloys and the alloy with a Pt : Cr ratio of 75 : 25 maintains its small advantage over an extended period.

Platinum usage is minimized by dispersing the catalyst over a high surface area support. This is usually a carbon black with a degree of graphitic character and a high surface area ($>75\,m^2\,g^{-1}$); typical carbons are Vulcan XC72R, Ketjen Black and Shawinigan Black. The platinum can be deposited by several procedures. One major manufacturer of Pt electrocatalysts forms a slurry of the carbon particles in an aqueous solution of a platinum salt and then adds a chemical reducing agent. Other approaches include impregnation of the carbon with a platinum salt, adsorption onto the carbon surface or ion exchange of surface sites with a platinum ion; this initial step is then followed by gas-phase reduction with H_2 at elevated temperature. The objective is to form a catalyst with $\sim 40\%$ by weight Pt on the carbon. The platinum is highly dispersed over the carbon surface and studies by X-ray diffraction show that a 40% by weight Pt on carbon catalyst prepared by chemical reduction has Pt centres with an average size of 2.2 nm and a surface area of Pt determined electrochemically of $120\,m^2\,g^{-1}$. Increasing the % of Pt leads to larger centres with less surface area per gram of Pt. On the other hand, centres $<2\,nm$ in size lose their catalytic activity since they are clusters of a small number of Pt atoms and, consequently, do not have the same metallic properties.

Other approaches to reducing the use of platinum are under investigation in the laboratory. For example, catalysts have been prepared where centres of a cheaper precious metal such as Pd are deposited and then coated with a monolayer or a few layers of platinum.

8.4.3   Anode Catalyst

The hydrogen oxidation reaction, Equation (8.13), is a facile reaction of many surfaces and, even on smooth Pt, the exchange current density can be as high as 10^{-3} A cm^{-2}. In consequence, the overpotential is negligible compared to that at the O_2 cathode. The catalyst for the anode in a PEM fuel cell must, however, be a precious metal because cheaper metals corrode in the acid environment. Platinum is the metal most widely used. The mechanism of the reaction is thought to take place by disassociation of the hydrogen molecule into adsorbed hydrogen atoms, followed by oxidation (Chapter 5).

The major issue with the anode catalyst under the operating conditions of the PEM fuel cell is poisoning by carbon monoxide. With pure Pt catalyst, the CO content of the H_2 feed must be below 10 ppm and this is difficult to achieve if the hydrogen comes from reforming of a carbonaceous fuel. Platinum/ruthenium alloys are more tolerant to CO poisoning because surface oxidizing agents are more readily formed on ruthenium than on platinum [*cf.* Equations (5.37) and (5.38)] and the mechanism for the oxidation of CO on the alloy becomes:

$$Ru + H_2O - e^- \longrightarrow Ru-OH + H^+ \tag{8.21}$$

$$Ru-OH + Pt - CO \longrightarrow \xrightarrow{-e^-} CO_2 + Ru + Pt + H^+ \tag{8.22}$$

A second factor is that the Pt–CO bond is weaker with the PtRu alloy than with pure Pt. When CO poisoning is an issue, the anode catalyst is commonly $Pt_{0.5}Ru_{0.5}$ although ternary alloys including metals such as Mo or W have been reported to improve performance further. Another approach is to bleed into the anode gas feed a low level of air ($<2\%$) and to include in the anode structure, between the electrocatalyst layer and the gas diffusion layer (see below), a thin layer of Pt/C catalyst. This acts as an efficient heterogeneous catalyst for the gas phase reaction of CO with O_2.

The anode catalyst can be damaged by cell current reversal. This is particularly common in large cell stacks and multi-stack systems and most commonly arises from reactant starvation, either O_2 at the cathode or H_2 at the anode, *e.g.* as a result of blockage or water flooding of a flow field channel or an instantaneous increase in power demand from the fuel cell. In a bipolar stack, every cell must pass the full current and if there is insufficient reactant, water electrolysis must occur. Oxygen will be evolved at a fuel starved anode and hydrogen at a cathode receiving insufficient oxygen. This will lead to an immediate drop in the stack voltage (hence, power output) but also long-term damage to the anode catalyst because corrosion of the carbon substrate occurs after only a few minutes at the more positive potential required for oxygen evolution at the current density drawn from a PEM fuel cell (often >0.5 A cm^{-2}).

8.4.4   Membrane Electrode Assemblies

The membrane electrode assembly consists of the anode, membrane of proton-conducting polymer and cathode integrated into a single component. In addition to the performance of the membrane and catalysts alone, fabrication of the MEA must lead to an effective, high area interface between the polymer and catalysts and high fluxes of gaseous reactants to the electrocatalyst/polymer interfaces.

MEAs have been fabricated by bonding a membrane and two pre-manufactured gas diffusion electrodes using compression at elevated temperatures. Generally, however, they are made by an entirely different approach. Notably, in a PEM fuel cell, there is no need for hydrophobic layers to resist electrolyte penetration through the catalyst layer and the contact area between the catalyst and electrolyte is determined by quite different factors. Most commonly, the anode and cathode layers are applied directly to the membrane surfaces (usually still Nafion™), by preparing an ink and then painting with a brush (in the laboratory) or by spraying or screen printing (on a larger scale). The ink is composed of the dispersed catalyst on high area carbon powder, a soluble form of Nafion™ (in water or an alcohol) and solvent. The catalyst in the cathode ink will typically be 40% Pt by weight on carbon powder (Section 8.4.2) and, in the anode ink, 20% Pt + 10% by weight Ru on carbon powder (Section 8.4.3). The role of the soluble Nafion is to increase the area of interface between the catalyst and polymer electrolyte in the MEA. The correct rheology of the ink is critical in obtaining uniform layers. The catalyst layers may be applied in single or multiple layers with drying between applications until one has the desired catalyst loading. The anode catalyst layers are thinner and the final Pt loading is $\sim 0.1 \, \text{mg cm}^{-2}$ of geometric area of the membrane/ electrode interface; for a typical dispersed Pt on carbon electrocatalyst, this will give $\sim 120 \, \text{cm}^2$ of Pt surface for each $1 \, \text{cm}^2$ of MEA. The cathode catalyst layer is somewhat thicker and has a final Pt loading of $\sim 0.25 \, \text{mg cm}^{-2}$ of geometric area of the membrane/electrode interface. The final stage of MEA fabrication is to compress the layers at elevated temperature and during this step a gas dispersion layer (*e.g.* a carbon paper) may be included behind each catalyst layer to aid the distribution of gas, especially in areas where the electrode is in contact with the bipolar plate. The whole MEA will have a thickness $< 1 \, \text{mm}$.

The electrochemical performance of the MEA will depend strongly on the operating conditions of the fuel cell (*e.g.* purity of H_2) but also on the overall design and the fabrication procedure as well as the selection of materials. A low membrane resistance, the optimum size of the catalyst centre, a high interfacial area between catalyst and polymer electrolyte, the efficient supply of gases to this interface and the control of the water content throughout the MEA are all critically important.

8.4.5 Bipolar Plate

Each individual cell will consist of the MEA and gaseous reactant supply systems. It is also necessary to control the humidities of the two gases to maintain the water content of the membrane at the optimum level and to remove heat from the cell as the inefficiencies (overpotentials, *IR* drops) are all converted into heat. In a stack, see Figure 8.4, there must also be a barrier to mixing of gases between O_2 and H_2 in neighbouring cells but very good electrical contact between cells. The bipolar plate between neighbouring cells has an important role in meeting several of these requirements.

The bipolar plate is fabricated from a material that is a good electrical conductor (to minimize *IR* drop between neighbouring cells) and also a good heat conductor to remove the heat from the stack. The material of construction must also be stable to corrosion at the operating temperature in an environment that is moist, oxidizing (oxygen atmosphere) and perhaps acidic (through exposure to Nafion™). While stainless steel and some other coated metals may be possible, the standard material is carbon. Into both surfaces is machined a convoluted channel (maybe

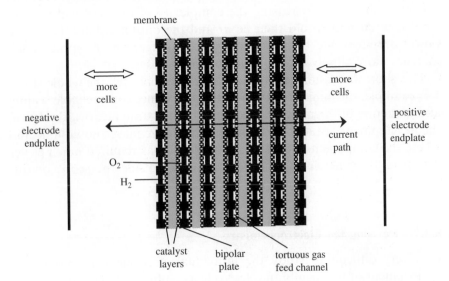

Figure 8.4 Conceptual figure of a stack of PEM fuels cells.

$1 \times 1\,mm$) pattern that is designed to ensure a uniform supply of reactant gas over the back of the whole catalyst layer. The channel on one side will be fed with H_2 and that on the other with O_2 or air; either or both gaseous feeds will have controlled additions of water vapour. The channels may also be used to remove excess water.

8.4.6 Fuel Cell Stack

The power that can be delivered from a single fuel cell is limited; for example, a power output of $1\,kW$ would require an MEA with dimensions of approx. $40 \times 40\,cm$. Hence, there is a requirement to consider ways to employ multiple cells. The usual answer is to develop a fuel cell stack. In a bipolar stack, a large number, anywhere between 6 and 500 cells, is sandwiched between two end plates that serve as the only electrical contacts to the stack (Figure 8.4). The cells are therefore electrically in series but in parallel with respect to the gas flows. The cell current will pass sequentially through each of the cells and will depend of the area of the MEAs and the current density, *i.e.* $I = jA$. The stack voltage will be nV_{cell} where n is the number of cells in the stack and V_{cell} is given by Equation (8.2); the power output will be nIV_{cell} watts. Here, the *IR* drop term will include the voltage drop through the membrane, catalyst layers and the bipolar plate under the operating conditions. In principle, it would be possible to operate the cells with a parallel electrical connection; this, however, requires two electrical connections to each cell, greatly increasing the complexity of the electrical system. The stack current would then be nI and the output voltage V_{cell}. The power output would be the same but the voltage from a bipolar stack matches more closely the requirements of the user.

   This short discussion of a fuel cell stack does not tell the whole story. For example, the miniaturization of the gas connections, reliable sealing and gasketing of the many components are not straightforward. Hence, the development of fuel cell stacks that minimize mass and volume and where the components can be mass produced has required much clever design and engineering. This, however, lies outside the scope of this book.

8.4.7 Revising the Electrochemistry

It is very difficult to study the individual electrode reactions in the environment of the membrane electrode assembly (MEA); as a result, it is more normal to present plots of current density *versus* cell voltage

(Section 8.4.8). It is, however, valuable to give some thought to the two individual electrode reactions.

As always, each of the electrode reactions is a sequence of two steps, supply of the reactant to the site of electron transfer followed by the electrode reaction. In the situation of a MEA, however, transport is considered to be occurring in the gas phase and the flux of reactant to the surface will be much larger than can be achieved in solution because the diffusion coefficient will be much larger. In practice, with good MEA design, the limiting current density will be $> 1 \, A \, cm^{-2}$ but, even so, there will be a mass transport limited current and this will depend on the structure of the MEA (*e.g.* the porosity of any gas dispersion layer and structure/composition the catalyst layer). At current densities well below the limiting current density, it is to be expected that the reactions are electron-transfer controlled and the current density will increase exponentially with overpotential.

Figure 8.5(a) and (b) shows data for the oxygen cathode. In both parts, the curves labelled 'a' are, in fact, for an oxygen fed GDE with a Pt loading of $0.5 \, mg \, cm^{-2}$ in 1 M H_2SO_4 at 353 K. The current increases smoothly with overpotential and, indeed, the data could be replotted to give a linear log j vs E plot with a Tafel slope of $1/70 \, mV$. In this figure, the current density is calculated based on the geometric area of the electrode. Notably, those interested in the fundamentals of electro-catalysis will prefer to calculate the current density based on the real electrochemical surface area estimated from the charge under the hydrogen adsorption peaks on a cyclic voltammogram. In contrast, a constructor of fuel cells would discuss the data in terms of amperes/gram of Pt since this represents the way to estimate the cost of the device

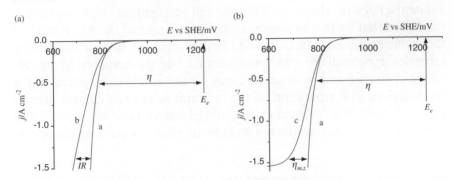

Figure 8.5 The oxygen cathode. Voltammograms to show the influence of (*a*) over-potential for reduction and *IR* through the cell and (*b*) the overpotentials for reduction and mass transport on the performance of the O_2 cathode in a PEM fuel cell.

delivering the desired current. As expected, the charge transfer over-potential is large, $>400\,\text{mV}$ to deliver $0.25\,\text{A}\,\text{cm}^{-2}$. Figure 8.5(a) shows the additional loss in cell voltage associated with half the *IR* drop in a typical PEM fuel cell; it increases linearly with current density but is always small compared with the charge transfer overpotential; the value is only $30\,\text{mV}$ with a current density of $0.5\,\text{A}\,\text{cm}^{-2}$. Figure 8.5(b) illustrates the influence of mass transport on the response; the curve labelled 'c' is drawn for a limiting current density of $1.6\,\text{A}\,\text{cm}^{-2}$. The effect of mass transport is to introduce a further cell voltage loss, the mass transfer overpotential; this is small at low current densities but becomes substantial as the limiting current density is approached. As noted above, the actual value of the limiting current density will depend on the structure of the MEA but with a pure oxygen feed, values in the range $1–10\,\text{A}\,\text{cm}^{-2}$ are common with the operating conditions of a PEM fuel cell. Of course, with an air feed, this limiting current will be reduced by a factor of almost five.

The charge transfer overpotential at the hydrogen anode is much lower due to both a higher exchange current density (by a factor of $>10^6$) and a steeper Tafel slope; the value is typically $<50\,\text{mV}$ at $0.25\,\text{A}\,\text{cm}^{-2}$. There will again be a mass transport overpotential and one could also assign half the cell *IR* to this electrode. The diffusion coefficient for H_2 in hydrogen will be significantly larger than that for O_2 in oxygen; the limiting current density will be higher and the mass transfer overpotential will be correspondingly less.

Figure 8.6 shows current density *vs* potential plots for both hydrogen oxidation and oxygen reduction under conditions similar to those found in an H_2/O_2 PEM fuel cell. Several features should be noted. In a device generating electrical energy, the cathode is the positive electrode and the anode the negative electrode. Also, the cell voltage that is achieved is less than predicted by thermodynamics and in the case of the PEM fuel cell the inefficiencies are substantial. At both electrodes, the current density increases exponentially with overpotential but the response at the H_2 anode is steeper and is observed closer to the equilibrium potential for the reaction. The separation of the two curves at each current density gives a simple way to estimate the likely cell voltage for the fuel cell although this estimate would not include the *IR* losses in the actual device.

8.4.8   PEM Fuel Cell Performance

The electrochemical performance of the PEM fuel cell is usually dis-cussed in terms of the energy efficiency and the power density that can be

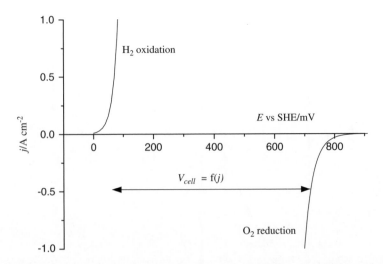

Figure 8.6   Current density *vs* potential plots for hydrogen oxidation and oxygen reduction under fuel cell conditions. The difference in the potentials for the same current density for each electrode reaction allows an estimate of the fuel cell voltage at that current density (it does not allow for *IR* drop in the cell).

delivered. Figure 8.7 shows a typical current density *vs* cell voltage characteristic for a PEM fuel cell fed with pure H_2 and pure O_2. As shown, current densities up to $1 \, A \, cm^{-2}$ can be delivered with a cell voltage $>650 \, mV$. Hence, the energy efficiency is in the range 50–60%, depending on the current density. Figure 8.8 shows the same data converted into a plot of power output as a function of cell voltage. The power output increases rapidly with increasing current density but then passes through a shallow maximum before falling off rapidly when mass transport influences the current density significantly; in the data of Figure 8.8, the influence of mass transport has been minimized by using an O_2 feed to the cathode. The maximum power density is $\sim 0.8 \, W \, cm^{-2}$ and this occurs at rather high current density where the energy efficiency will be lower than for lower current densities; the selection of conditions will be a compromise between the power output and energy efficiency.

Key advantages of the PEM fuel cell include a high power density output compared to batteries and other fuel cells as well as their compactness. However, the energy efficiency is only moderate. This, of course, results almost entirely from the high overpotential for oxygen reduction and it has to be recognized that there are presently no leads to the development of a substantially improved electrocatalyst for this

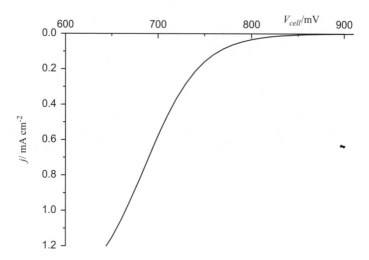

Figure 8.7   Variation of cell voltage with current density for a PEM fuel cell fed with pure O_2 and pure H_2 at 353 K.

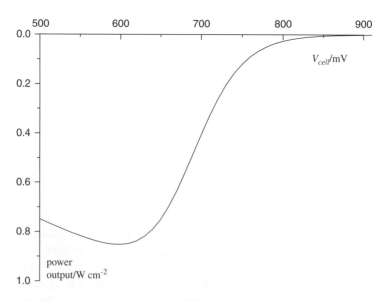

Figure 8.8   Power output as a function of cell voltage for a PEM fuel with the current density *vs* cell voltage characteristic shown in Figure 8.7.

reaction. Hence, today's R & D is focused on smaller increment improvements such as:

- reducing the loading of the expensive electrocatalysts without adversely influencing the current density *vs* cell potential characteristic;

- maintaining the initial performance over the whole operational life of a fuel cell system, maybe several years;
- developing low cost (*i.e.* non-fluorinated) polymer membranes and seeking the next generation of membranes that can be operated at up to 410 K (to enhance the O_2 reduction kinetics and to decrease the influence of CO poisoning of the anode catalyst);
- seeking improved stack design and lower cost procedures for mass production.

8.4.9 Commercial Developments

PEM fuel cell stacks have been developed for electricity generation on a large scale. Units with up to 200 kW power output have been operated by electricity supply utilities while smaller PEM fuel cell units with power outputs up to 25 kW have acted as stationary power supplies to the military, to remote sites, hotels, hospitals and houses. Mostly such units will operate alongside a reformer unit fed by methanol or natural gas and be part of a combined heat and power system so that the total energy efficiency approaches 90%.

The biggest market for PEM fuel cells is thought to be transportation. Most of the large automobile manufacturers have developed an electric car with a PEM fuel cell powering the engine. Perhaps the nearest to market are Toyota, Honda, GM and Daimler Chrysler. Figure 8.9 shows pictures of the GM vehicle and its PEM fuel cell stack. The Toyota FCVH4 is said to have a similar performance to a petrol engine car and it has an overall energy efficiency of 48% compared with 16% for a modern petrol powered car. The problem is the absence of a distribution network for hydrogen and the cost of the fuel cell stack. Presently, a fuel cell powered car is more expensive than a petrol-engine vehicle by a factor of >2 although that differential would drop with mass production.

Ballard has focused on fuel cell stacks for buses where a H_2 production facility could be sited at the depot. Buses are operating in nine European cities as well as several North American cities (including Santa Clara, Chicago, Washington and Vancouver), Australia and China. All have performed well in service, with minimal emissions, smell and noise and good acceptance by both passengers and operators. The 6th generation heavy duty fuel cell modules for buses are rated at 75 or 150 kW (325 A at 220 or 440 V) and are housed in a unit $129 \times 89 \times 49$ cm weighing <350 kg.

(a)

(b)

Figure 8.9 (*a*) A car powered by a PEM fuel cell stack. (*b*) The fuel cell stack. (Photographs are by courtesy of General Motors.)

Smaller PEM fuel cell systems with power outputs from a few kW down to a few W are also being developed to power, for example, a mobile home. The smallest units, designed perhaps to replace a battery in powering a mobile phone or laptop computer, may be based on the direct oxidation of an organic fuel, with methanol or formic acid as the most likely fuels. Such PEM fuel cells need, however, larger catalyst loadings.

Numerous companies claim to be close to marketing PEM fuel cells. All of the systems delivered so far, however, must be considered only as demonstration units since most are subsidized by government or other agencies with a view to confirming their potential for providing reliable power in an environmentally friendly way. Regrettably, PEM fuel cells, along with other types of fuel cell, would still not be competitive on a purely economic basis and their installation relies on their substantial environmental advantages.

Further Reading

1. *Handbook of Fuel Cells, Volumes 1–4*, ed. W. Vielstich, A. Lamm and H. A. Gasteiger, John Wiley & Sons, New York, 2003.
2. *Fuel Cells Technology Handbook*, ed. G. Hoogers, CRC Press, Boca Raton, FL, 2003.
3. A useful website: www.fuelcells.org, Jan. 2009.
4. T. R. Ralph and M. P. Hogarth, *Platinum Met. Rev.*, 2002, **46**, 3–117.
5. V. Mehta and J. Smith Cooper, *J. Power Sources*, 2003, **114**, 32.
6. T. A. Davis, J. D. Genders and D. Pletcher, *A First Course in Ion Permeable Membranes*, The Electrochemical Consultancy, Romsey, 1997.
7. J. C. Larminie and A. Dicks, *Fuel Cell Systems Explained*, John Wiley, New York, 2003.
8. A. Wieckowski, *Fuel Cell Catalysis: A Surface Science Approach*, John Wiley, New York, 2009.

CHAPTER 9

Improving the Environment

9.1 INTRODUCTION

At the very beginning of this book (Table 1.1) the many, diverse applications of electrochemistry were highlighted and the details of these technologies can be found in the texts listed in Further Reading at the end of this chapter. Electrochemical technology can contribute in many ways to the development of a world with a cleaner and more stable environment. These include:

- providing renewable energy to the consumer and for transport;
- reducing the outflow of CO_2 and other greenhouse gases to the atmosphere;
- using natural resources (fuels, chemical feedstocks, minerals, ores, water) more efficiently;
- bringing about chemical change, cleanly in safe conditions;
- removing toxic and otherwise harmful materials from industrial effluents;
- improving the quality of water for drinking and other uses.

Some of these possibilities will be illustrated in this chapter. The approach will be not to attempt complete coverage but rather to provide 'case studies' to show how knowledge of electrode reactions can be converted into technology for the modern world and, indeed, the future. A unifying feature is that all the applications will use a voltage applied between two electrodes to drive chemical change that otherwise would

A First Course in Electrode Processes, 2nd Edition
By Derek Pletcher
© Derek Pletcher 2009
Published by the Royal Society of Chemistry, www.rsc.org

not take place (the overall cell reaction has a positive Gibbs free energy). This contrasts with a fuel cell (discussed in Chapter 8) where a spontaneous overall cell reaction is used to create electrical energy.

All technology has a specific goal that must be achieved at an acceptable cost. In electrochemical technology, the economics of electrolysis are usually dominated by two factors:

1. The investment in the electrolytic cells, both the initial cost (dependent on the design and materials of construction) and that of replacement components (dependent on their operational lifetime). Always, the initial cost of the cells is inversely proportional to the current density for the process. Hence, the drive to operate at the highest possible current density, perhaps even under mass transport control.
2. The annual cost of energy and hence the drive to minimize the cell voltage used to drive the cell chemistry. This implies minimizing the overpotentials for the electrode reactions and certainly operating under electron-transfer control as well as minimizing IR losses.

The relative importance of these two factors always depends on the nature of the technology and almost always on the size of the electrolysis system/plant, with energy consumption becoming more important as the size of the plant increases and, in consequence, the amount of energy consumed. Also, it can be seen that the dominance of one of the two factors will lead to different approaches to process design; electron-transfer control places the emphasis on the development of electrocatalysts, while mass-transfer control will require attention to the mass transport regime within the cell. In both situations, there is an advantage in increasing the electrode area/volume of the cell but the approach to increasing the electrode area will be different (essentially the scale of the surface roughness). In many other processes, where neither the cost of equipment nor energy is totally dominant, there is a need for compromise between the factors tending to electron-transfer or mass-transfer control in order to obtain the optimum cell performance. It also needs to be recognized that the different 'goals' of the technologies discussed in this chapter may put emphasis on quite different criteria, *e.g.* full conversion of reactant in a single pass of the reactant through the cell or high selectivity for a particular electrode reaction. Finally, in the real world, the electrolysis cell is usually only one part of a complex group of interacting unit processes constituting the process, and optimizing the 'process performance' may lead to operating the electrolysis cell away from its optimum conditions.

9.2   WATER ELECTROLYSIS

On various scales, water electrolysis is used to produce clean oxygen and/or clean hydrogen. For example, it has been used to produce clean oxygen at remote hospitals and for the atmosphere in submarines. Also, water electrolysis has been used as a local source of clean hydrogen for the manufacture of electronic components and in the laboratory for gas-phase chromatography detectors. As noted in the previous chapter, in the future, water electrolysis (alongside H_2/O_2 fuel cells) may be used on a much larger scale to store energy from renewable sources for supply to the consumer at another time. The overall cell reaction is:

$$2H_2O \longrightarrow O_2 + 2H_2 \tag{9.1}$$

and the cell voltage to drive this reaction is given by:

$$-|V_{cell}| = |\Delta E_e| + |\eta_a| + |\eta_c| + |IR| \tag{9.2}$$

where the thermodynamic driving force, ΔE_e, is 1.23 V and the other terms are inefficiencies that increase the cell voltage and hence the energy consumption and lead to heat; these terms should therefore be minimized. Since the overall cell reaction has a positive Gibbs free energy, the cell voltage will be negative. Equation (9.2) is written in a way to emphasize that the objective is always to minimize the magnitude of the (negative) cell voltage. The importance of energy efficiency, here defined as:

$$\text{Energy efficiency} = \frac{\Delta E_e}{V_{cell}} \cdot 100 \tag{9.3}$$

depends on the size of the system and hence the amount of energy that it consumes. With small units, the energy efficiency will not be a major factor in their economics. With a large system such as technology for energy storage, the energy efficiency will be a dominant factor determining whether it is successful.

Many water electrolysers employ an alkaline electrolyte since this allows the use of non-precious metal electrocatalysts. Then, the electrode reactions are:

$$2H_2O + 2e^- \longrightarrow H_2 + 2OH^- \tag{9.4}$$

$$4OH^- - 4e^- \longrightarrow O_2 + 2H_2O \tag{9.5}$$

The water electrolyser may be configured as a simple tank cell or a parallel plate, filter press cell; Figure 9.1(a) shows a sketch of the key components. The separator has the role only of keeping apart the O_2

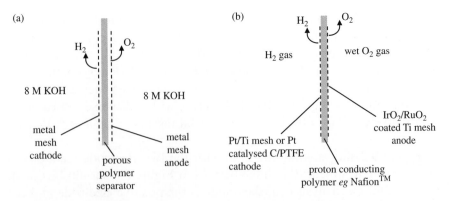

Figure 9.1 Schematic diagrams of water electrolysers based on (*a*) an aqueous alkaline electrolyte and (*b*) an acidic solid polymer electrolyte.

and H$_2$ products and can therefore be fabricated from several porous polymeric materials. The electrolyte is ∼8 M KOH as this gives the highest conductivity and the cells are usually operated at ∼353 K to increase the conductivity further and reduce the overpotentials associated with the electrode reactions, particularly O$_2$ evolution. The electrodes are usually pushed up to the surface of the separator to minimize the interelectrode gap and are meshes so that the gaseous products can escape from the interelectrode gap; both are designed to keep the *IR* drop as small as possible. Typically, the anode material will be a high surface area form of Ni plated onto to steel or Ni/Co spinel coated Ni or steel and the cathode a high surface area Ni or Ni alloy on a nickel or steel substrate. Such alkaline water electrolysers are usually operated with a current density of 0.1–0.6 A cm^{-2}, requiring a cell voltage of −1.9 to −2.6 V. The major inefficiency is the overpotential associated with the O$_2$ anode.

The competing technology is based on a solid polymer electrolyte, usually the perfluorinated sulfonic acid, Nafion™. Under acid electrolyte, the electrode reactions become:

$$2H^+ + 2e^- \longrightarrow H_2 \tag{9.6}$$

$$2H_2O - 4e^- \longrightarrow O_2 + 4H^+ \tag{9.7}$$

and the electrodes are pressed up against the membrane to make electrical contact (Figure 9.1b). With the conducting polymer electrolyte, the only requirement for mass balance is to feed water to the membrane, usually as vapour in the anode compartment. In the acidic conditions, for stability to corrosion, the electrocatalysts are precious metals. Since the potential for oxygen evolution (~ 1.6 V) is substantially positive to that for oxygen reduction in a fuel cell (~ 0.7 V) it is not possible to use a carbon substrate (the carbon corrodes rapidly) and so a different electrode structure must be employed. The anode catalyst is therefore an IrO_2/RuO_2 coating on a fine Ti mesh support. The cathode may be similar to a fuel cell negative electrode and based on a C/PTFE structure with a dispersed Pt catalyst. Alternatively, it may also be a coated metal mesh, *e.g.* high area Pt on a Ti mesh.

Figure 9.2 shows voltammograms for the two electrode reactions in an acidic environment; the differences from Figure 8.6 should be noted. In the water electrolysis cell, the anode is the positive electrode and the cathode the negative electrode. Again, the current densities for the two electrode reactions increase exponentially with overpotential as the reactions will be under electron-transfer control but now the cell voltage is significantly more than that predicted by thermodynamics as the inefficiencies add to the cell voltage (with fuel cell chemistry the inefficiencies decrease the cell voltage).

Solid polymer electrolyte cells can be operated with a current density of ~ 1 A cm^{-2} with a cell voltage of *ca.* -2.0 V, which is significantly more energy efficient than the alkaline fuel cells. Hence they are likely to

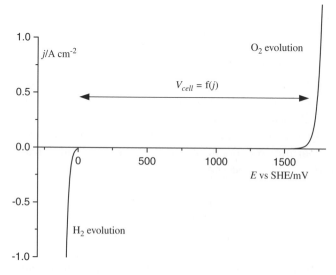

Figure 9.2   Current density *vs* potential plots for hydrogen evolution and oxygen evolution under the conditions of a water electrolysis cell with an acidic environment. The difference in the potentials for the same current density for each electrode reaction allows an estimate of the cell voltage at that current density (it does not allow for *IR* drop in the cell).

be the basis of new large facilities. The major inefficiency remains the overpotential associated with the O_2 anode.

9.3 PROVIDING CLEAN WATER

High quality water is essential for humans. Although natural waters can meet the standard required, it is more common for it to require treatment before use for drinking, cooking, *etc.* This treatment may involve the removal of bacteria and salts as well as inorganic and organic materials, often introduced by industrial processes or agricultural activities. Electrochemical technology contributes substantially to the available procedures. For example:

- chlorine, the most widely used chemical in the water industry, is manufactured by electrolysis;
- where natural waters contain high levels of salts, electrodialysis is one method for their removal;
- several electrochemical approaches have been developed for the removal of nitrate ion, now frequently present in natural waters because of run off from agricultural land resulting from its use as a fertilizer;
- several electrochemical methods for the removal of heavy and transition metals are available commercially; these include technology based on three-dimensional electrodes (Section 9.5);
- likewise electrochemical methods can be used for the removal of organics; these include both direct anodic oxidation and indirect electrolyses where an appropriate redox reagent such as hydrogen peroxide, hypochlorous acid or ozone is produced in the cell and used for decontamination of the water in a different reactor.

Electrochemical methods work best when the 'reactant' is present in 'high' concentration. Hence, wherever possible, technology for the removal of toxic materials should be operated at the site where they are produced, *i.e.* as effluent treatment processes (Section 9.5).

Here, technologies to improve the quality of water will be illustrated by focusing on units for the *in situ*, on demand production of chlorine (or hypochlorous acid or hypochlorite). It was noted above that chlorine remains a preferred chemical for killing off bacteria within water utilities. In addition, it is widely used for sterilizing instruments in hospitals, for protecting foodstuff from contamination by bacteria, the conditioning of swimming pools, treating cooling tower water, *etc.* The transport and handling of chlorine gas is expensive and requires care as it is a highly hazardous chemical. Therefore, unsurprisingly, several companies market units for on-site, on-demand generation of chlorine or chlorine in soluble forms. The performance of these electrolysis systems benefits greatly from knowledge gained within the chlor-alkali industry as well as cells and components developed for this industry. The chlor-alkali industry:

- manufactures >40 million tons of chlorine and sodium hydroxide each year at many sites throughout the world;
- passes continuously a current of $\sim 4 \times 10^9$ A, employing electrode areas $> 10^6$ m^2;
- consumes $\sim 10^{13}$ Wh of electricity per year (~ 1–2% of all the electricity produced).
- For an industry of this size, an intensive effort inevitably goes into cell design and component development to reduce costs, including minimizing the energy consumption.

The anode reaction in these cells is:

$$2Cl^- - 2e^- \longrightarrow Cl_2 \tag{9.8}$$

with an equilibrium potential of ~ 1.32 *vs* SHE. Hence, it should immediately be noted that oxygen evolution ($E_e \approx 1.23$ V *vs* SHE at pH 0 and less positive at higher pH) is always the thermodynamically preferred reaction. The evolution of chlorine is dependent upon the kinetics of Cl_2 evolution being fast while O_2 evolution is hindered. In fact, chlorine is stable only in acidic solutions. Otherwise, chlorine is hydrolysed and, in neutral solution, hypochlorous acid is the major species formed:

$$Cl_2 + OH^- \longrightarrow HOCl + Cl^- \tag{9.9}$$

while at pH 9 and above, hypochlorite is the dominant species:

$$Cl_2 + 2OH^- \longrightarrow OCl^- + Cl^- + H_2O \tag{9.10}$$

All three products are useful for sterilization processes and, anyway, the hydrolysis can be reversed by addition of acid. Hydroxide normally results from the cathode reaction:

$$2H_2O + 2e^- \longrightarrow H_2 + 2OH^- \tag{9.11}$$

The product actually formed therefore depends on the way the electrolysis cell is configured. If the cell is to deliver gaseous chlorine,

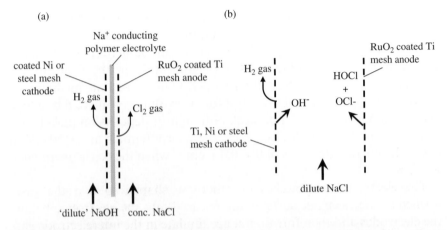

Figure 9.3   Schematic diagrams of electrolysis cells used in sterilization procedures
and the purification of water: (*a*) cell for the on-site, on-demand generation of chlorine gas and (*b*) cell for generating hypochlorous acid/
hypochlorite.

it will be a scaled down chlor-alkali unit (Figure 9.3a). The cell will be a
membrane cell with parallel electrodes in a narrow gap or zero gap
configuration, *i.e.* the electrodes will be sited close to the membrane or
actually touching the membrane surface.

In systems to generate gaseous chlorine, a membrane such as a
Nafion™ 900 series membrane with a bilayer structure is employed.
Most of the membrane is fabricated from a perfluorinated polymer with
sulfonate fixed ionic groups (Figure 8.3) but there is a thin layer of a
similar perfluorinated polymer, but with carboxylate fixed ionic groups,
on the cathode side. This carboxylate layer reduces the back migration
of hydroxide through the membrane: note that as well as a voltage field
driving the migration of Na^+ from anolyte to catholyte, the same field
will seek to drive hydroxide in the opposite direction. The carboxylate
polymer is much less hydrophilic than the sulfonate polymer and forms
a superior barrier to the unwanted transport of hydroxide. A drawback
with these membranes is that they are easily poisoned by divalent ions
that interact with the fixed ionic sites more strongly than monovalent
ions such as Na^+. As a result, the brine feed must be free from ions such
as Mg^{2+} and Ca^{2+} and this requires pre-treatment by precipitation as
hydroxides/carbonates and then ion exchange.

The electrocatalysts will also be taken from the chlor-alkali industry.
The anode catalyst is based on RuO_2 (perhaps, also containing other
metal oxides) coated onto a Ti substrate by thermal decomposition of a
salt solution(s) in air. These materials are known as dimensionally stable
anodes (DSA) because of their great stability in NaCl solutions (service

life > 5 years); they also give a low overpotential (usually < 50 mV) and a high selectivity for Cl_2 *vs* O_2 evolution (often above 99%). This stability contrasts strongly with alternative anode materials such as Pt and carbon, both of which corrode significantly in the concentrated chloride environment. The cathodes operate in an alkaline medium, see Figure 9.3(a), and hence the catalyst for H_2 evolution will often be a high area Ni deposited onto Ni or steel, although Pt deposits on nickel have also been used. At both electrodes, the overpotentials are < 50 mV at operating current densities of 0.1–0.4 A cm^{-2} when the cell is operated at ~ 353 K.

The electrodes are meshes or otherwise shaped so that the gases formed at the electrode surfaces are released into the electrolyte behind the electrodes and therefore do not accumulate in the interelectrode gap, thereby increasing the cell resistance and energy consumption. The electrolytes will be circulated through the cell from reservoirs. The anolyte will be concentrated NaCl (up to 25%) with addition of acid to maintain the pH in the range 2–3. The catholyte feed will be dilute NaOH and this will concentrate with time. In this alkaline medium, the equilibrium potential for hydrogen evolution will be *ca.* -0.84 V *vs* SHE. Hence, the contribution of thermodynamics to the cell voltage (ΔE_e) will ~ 2.16 V. The protocol for operating such on-site, on-demand units depends strongly on the size of the unit and the circumstance of where it is sited (*e.g.* whether it is within an industrial complex or isolated in the countryside). In larger scale plants, the cells will be operated at ~ 353 K to produce 32% NaOH but in smaller units it may be convenient to operate at ambient temperature (increasing overpotentials slightly and also increasing the cell resistance) and accepting the increased energy consumption. Also, the cell operation will depend on the source and purity of the NaCl feedstock and how the sodium hydroxide generated will be used; a more dilute NaOH product stream may be preferred. Likewise, the fate of the hydrogen will depend on the scale of the plant and where it is sited. Ideally, it should be used in a chemical or manufacturing process but it may also be burnt or simply vented from small units. Typically, the cell will be operated with a current density of 0.1–0.5 A cm^{-2} with a cell voltage between -2.7 and -3.2 V.

Cells to produce the soluble species, hypochlorite and hypochlorous acid, generally produce a mixture because of the equilibrium:

$$ClO^- + H^+ \rightleftharpoons HClO \qquad (9.12)$$

and hence the product concentration is reported in terms of the total equivalent chlorine concentration. Soluble forms of chlorine tend to be more convenient to handle than chlorine gas, but when generated by electrolysis their solutions are almost always contaminated with unconverted NaCl. The systems are usually smaller and the cells are configured very differently from those intended to generate chlorine gas. Firstly, the cell can be undivided; the OH^- generated at the cathode is used to maintain the electrolyte pH constant and in the neutral medium the chlorine hydrolyses rapidly to give a dissolved product so that mixing of two gases is not a problem. Contact of the hypochlorous acid/ hypochlorite with the cathode must be limited since they may both be reduced and this factor usually limits the concentration of hypochlorous acid/hypochlorite that can be generated in an undivided configuration. In some technologies a separator has been put into the cell and the electrolyte fed from cathode to anode so that the chlorine is generated in an alkaline environment; this approach, however, makes the cell design more complex. Secondly, it is usually desirable to pass the solution only once through the cell so that the electrode area/cell volume and the residence time must be designed to allow the formation of the desired product concentration in a single pass. Thirdly, the cell feed can be variable. In some plants the cell feed is seawater (*i.e.* impure 2% NaCl) with the product being 1000–2500 ppm Cl_2 equivalent and used to prevent organic fouling in water intakes (to, for example, power station cooling systems) from the sea. In other units where high concentrations of NaCl in the product solution would be a disadvantage, the cell feed might be as dilute as 2–10 g dm^{-3} NaCl and the hypochlorite in the product solution 200–500 ppm Cl_2 equivalent. With a decrease in chloride concentration in the feed, the cell voltage must increase and the mass transport regime becomes more important (both increasing the energy consumption). Because energy consumption is less of an issue, robustness often determines the selection of electrode materials; anodes are almost always a RuO_2 coating on Ti while the cathode may be steel, nickel or titanium. With impure feeds (*e.g.* seawater or solutions prepared with food grade NaCl), a major problem is the deposition of magnesium and calcium hydroxides in the alkaline environment around the cathode. Some cell providers recommend periodic acid washing of the cell or even current reversal to create a local acidic environment to redissolve the hardness deposit. Several cell designs have been employed for the generation of soluble chlorine. An undivided parallel plate cell, Figure 9.3(b), can be used while a concentric pipe cell is a popular alternative.

Systems based on seawater are very common with many units around the world. They can also be rather large, with outputs up to 500 kg of Cl_2 equivalent per hour. Another widespread application is the provision of the sterilizing solution for swimming pools where the cells may produce $0.1–1\,kg$ of Cl_2 equivalent per hour. In other applications, the scale is again dependent on the local circumstances. Clearly, concern about energy consumption becomes less with decreased scale and the emphasis changes to reliable operation with unskilled operators and good quality product.

9.4 PRODUCTION OF FINE CHEMICALS

The fine chemicals industry produces numerous organic compounds, on scales from a few kilograms to 10000 tons per year. New markets open up regularly and there is a recognized need to replace chemical procedures that are inefficient in their use of feedstocks and/or use either toxic reagents or hazardous conditions.

For over 200 years, it has been recognized that electrolysis involves chemical change at each of the electrodes. Moreover, this concept has been used for the manufacture of chemicals, *e.g.* in the chlor-alkali industry and the extraction of aluminium, while an example of a large tonnage organic conversion is the hydrodimerization of acrylonitrile to adiponitrile as part of a route to nylon. In addition, a range of other electrolytic processes routinely manufacture both inorganic and organic compounds on a lower tonnage scale. This section considers the electrosynthesis of organic compounds for the fine chemicals industry since it is in this industry that the greatest opportunity for innovation arises as well as the opportunity to introduce new technology.

Electrolysis can contribute substantially to a fine chemicals industry that is kinder to the environment and employs safer plant. A major advantage of electrolysis in the manufacture of organic compounds is that it converts starting materials into products under mild conditions (usually at temperatures and pressures close to ambient) without the use of toxic and/or hazardous reagents. Indeed, in general, it avoids completely the use of stoichiometric reagents and therefore the certainty of large quantities of a by-product. This contrasts totally with an oxidation/reduction carried out with a redox reagent when separation of the organic product and the spent redox reagent is necessary and a strategy must be developed for the treatment of a waste stream often containing a high concentration of a toxic or otherwise hazardous element, *e.g.* Cr, Zn.

Numerous electrochemical oxidations and reductions have been reported in the literature. The challenge is to find conditions where the selectivity of the conversion is close to 100% so that the feedstock is used efficiently and minimal by-product streams are generated. With poly-functional substrates, it is in principle possible to control the potential of the electrode so that only a single functional group is oxidized or reduced. At the scale of production of fine chemicals, energy consumption is not normally an issue; the cost of the energy is usually small compared to the cost of the electrolytic flow cells and the chemical feedstocks. The electrode area necessary to form the desired quantity of product (and hence the cost of the cells) is determined by the current density for the reaction. It should again be noted that the limiting current density:

$$j_L = nFk_m c \qquad (9.13)$$

cannot be exceeded. A current density $50–250\,\text{mA cm}^{-2}$ is likely to be essential for an acceptable cell cost; such current densities will then lead only to the formation of $0.18–0.9\,\text{g cm}^{-2}\,\text{h}^{-1}$ ($n = 2$, molecular weight $= 200$). Such current densities are possible only if the organic substrate has sufficient solubility (perhaps $0.2–1\,\text{M}$) and there is an efficient mass transport regime in the cell. This, of course, limits the re-actions that can be operated by direct electrolysis. In consequence, several strategies (*e.g.* use of mixed solvents, indirect electrolysis, three-dimensional electrodes) have been developed to address this limitation and some have been successfully scaled-up. Another issue is the design of the technology to extract pure product from the electrolysis medium where the organic is present as a relatively dilute species along with electrolyte to decrease the cell voltage.

In many electrolytic processes, only the anode or cathode is used in the synthesis and the counter electrode chemistry is hydrogen or oxygen evolution. A clever strategy is to use both electrode reactions in a positive way provided the total cell reaction is balanced and both products can be sold and/or used on the scale that they are produced. This approach avoids even gaseous by-products from the counter electrode and splits both the investment and energy costs between two products. An example of such chemistry is employed by BASF GmbH in Germany. Figure 9.4 shows a reaction scheme for the production of phthalide and *t*-butylbenzaldehyde dimethyl acetal (later hydrolysed to the alde-hyde) from dimethyl phthalate and *t*-butyltoluene; both the products are used in downstream chemistry and are therefore required in equimolar amounts. It can also be seen that the overall cell chemistry is completely balanced with consumption of only the two reactants and production of only the two products. The electrolysis is carried out in a methanol based, organic medium using a simple, undivided cell developed within the company (Figure 9.5). The cell designis based on a stack of closely spaced carbon discs, $\sim 1\,\text{m}$ in diameter, separation $\sim 1\,\text{mm}$.

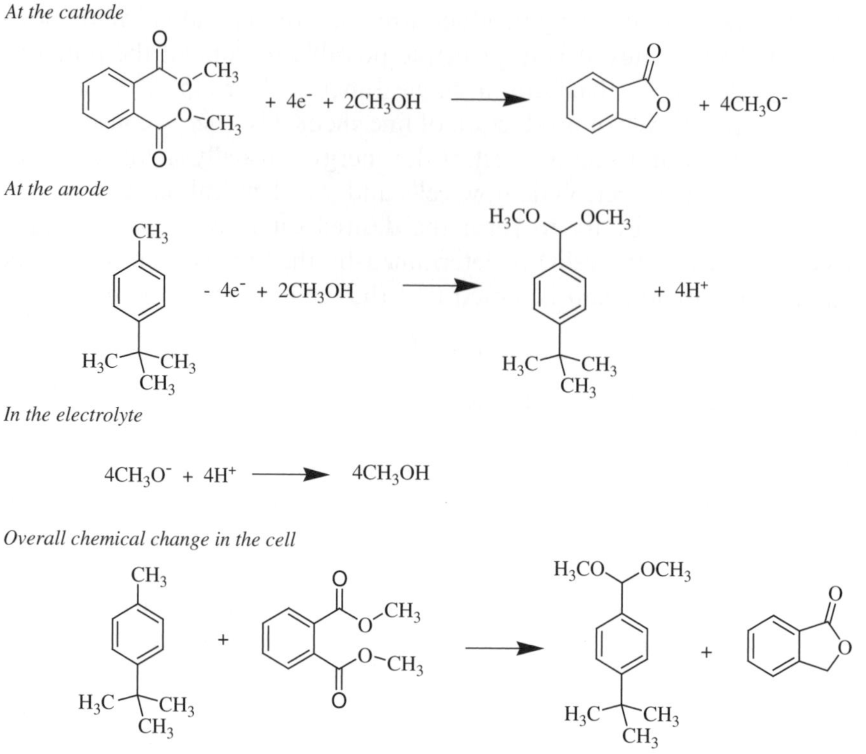

At the cathode

At the anode

In the electrolyte

$$4CH_3O^- + 4H^+ \longrightarrow 4CH_3OH$$

Overall chemical change in the cell

Figure 9.4 Chemistry of the BASF process for the conversion of dimethyl phthalate and *t*-butyltoluene into phthalide and *t*-butylbenzaldehyde dimethyl acetal, respectively.

The electrolyte solution is cycled through the narrow interelectrode gaps and the electrical connection is bipolar. The current density is $\sim 0.1\,A\,cm^{-2}$ and both the current yield and chemical yield approach 100%. Successful implementation of the process and integration into the overall technology required the development of a melt crystallization procedure to isolate pure products and the identification of an electrolyte, soluble in methanol and readily separated from the products.

In common with most high yield electrosyntheses, the intermediates in these reactions are very short lived and cyclic voltammograms of both reactants in the electrolyte medium would only show completely irreversible peaks. It is, however, possible to speculate on the mechanisms for the two-electrode reactions. Cyclic voltammograms to investigate the oxidation of larger aromatic molecules and polymethyl toluenes in non-nucleophilic media show a reversible $1e^-$ oxidation and it can be

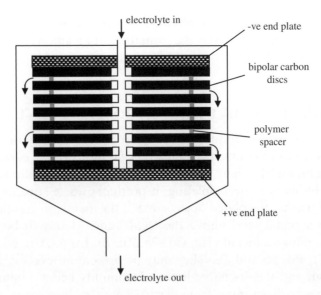

electrolyte in

-ve end plate

bipolar carbon discs

polymer spacer

+ve end plate

electrolyte out

Figure 9.5 BASF bipolar disc stack cell for the manufacture of fine chemicals by electrolysis.

concluded that the initial product is a cation radical of the reactant. The stability of the cation radical depends on the size of the delocalized electronic system and the medium but the trends would suggest that the *t*-butyltoluene is also first oxidized to a cation radical. Product studies of the oxidation of toluenes under various conditions show that products from both benzyl radicals and benzyl cations can be formed. Hence, it is likely that the reaction follows an ecec pathway involving loss of a proton from the cation radical before a further electron loss leads to a benzyl cation that reacts with the solvent:

$$\tag{9.14}$$

Indeed, under some conditions, it is possible to isolate the benzyl methyl ether as a product. In view of the formation of the diacetal in high yield, however, it is clear that the ether is more readily oxidized than the toluene and a further ecec sequence follows the formation of the ether. The mechanism of the cathode reaction is less certain but, again from studies of aromatic molecules in aprotic solvents, it can be postulated that the first step is the formation of an anion radical centred largely on the carbonyl group.

9.5 REMOVAL OF METAL IONS FROM EFFLUENT

Numerous industries produce effluent that contains heavy or transition metal ions. In most countries, there are legal limits to the concentration of each metal ion that may be discharged into sewers or natural waters. A typical industry is electroplating; it produces waste streams from the water that washes the plated objects free of the metal ion and this stream may have a metal level above the legal limit. Metals to be removed from such effluents include Hg, Cd, Pb, Zn, Bi, In, As, Cu, Ni, Cr, Mn, Ag, Au, Pt, Pd, Sb and Te. They may be present at levels in the range 1–1000 ppm and the discharge limit is commonly below 1 ppm. Several technologies with an input from electrochemistry have been developed to treat such effluents.

Here, we consider only those based on the cathodic reduction of the ions in the effluent to metal on the cathode:

$$M^{n^+} + ne^- \longrightarrow M \tag{9.15}$$

and most of the metals listed above can be removed by this reduction approach. Since the metal ions are present in the effluent only in low concentration, it will clearly be advantageous that the electrode reaction is carried out under conditions of mass transfer control. Moreover, the design of the electrolytic cell must have a very efficient mass transfer regime and most will also have a high surface area cathode. Again, because of the low charge requirement with dilute reactant, the energy consumption will not be an important issue. The challenge is to remove all the metal from the effluent and this can be realized by designing a cell that can achieve full conversion in a single pass or operating in batches by employing a system with a reservoir and recycling the effluent through the cell until the metal ion drops to the required level. In favourable circumstances, the value of the metal recovered can contribute substantially to the process costs.

In many effluents, the major competing reactions will be hydrogen evolution (either on the uncoated cathode material or, more likely, on the metal being deposited), oxygen reduction and any reducible organic contaminants. These reactions can be substantial charge consumers compared to metal removal reaction; the current efficiency for metal removal will depend on the relative concentrations of metal ions, oxygen and other reducible compounds in the effluent and the overpotential for hydrogen evolution. It should also be recognized that these metal removal technologies will operate best when the metal ion is uncomplexed (*e.g.* in acid media); complexing agents will cause the potential for the metal ion reduction to shift negative so that competing reactions are more likely to be serious interferences.

Several companies have built their electrolytic cells for effluent treatment around three-dimensional electrodes. Such electrodes combine a very high specific surface area (total surface area per unit volume) with efficient mass transport – the electrode material itself acts as a turbulence promoter. The limiting current density to a three-dimensional electrode must be written:

$$I_L = nFk_mA_eV_ec \qquad (9.16)$$

where A_e is the specific surface area and V_e is the volume of the electrode. Figure 9.6 shows two materials used for three-dimensional electrodes, reticulated vitreous carbon and carbon felt. The former is a very porous carbon foam with a relatively high specific surface area. Carbon felt has a much higher specific surface area but it is a much denser structure and therefore more difficult to pump liquids through it rapidly and uniformly.

Figure 9.7 illustrates the advantage of three-dimensional electrodes. It shows plots of limiting current *versus* mean linear flow rate for a solution containing 1 mM Cu^{2+}. In this 'parallel plate' cell, the plate cathode is 5×5 cm and the three-dimensional materials are blocks $5 \times 5 \times 1.2$ cm thick. It can be seen that the use of the 100 ppi reticulated vitreous carbon cathode scales the current by a factor of >100

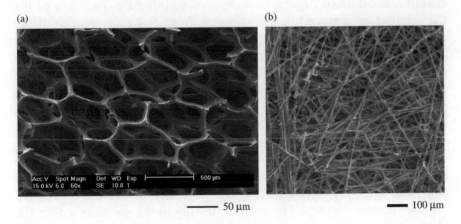

Figure 9.6 Scanning electron microscope images of (*a*) 60 pores/inch reticulated vitreous carbon and (*b*) carbon felt.

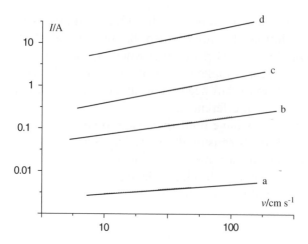

Figure 9.7 Variation of the current with the mean linear flow rate of a solution containing 1 mM Cu^{2+} + 0.5 M Na_2SO_4, pH 2 (deoxygenated) for cells with (a) a 5×5 cm flat plate cathode, (b) a $5 \times 5 \times 1.2$ cm, 10 ppi reticulated vitreous carbon cathode (specific surface area $8 cm^2 cm^{-3}$), (c) a $5 \times 5 \times 1.2$ cm, 100 ppi reticulated vitreous carbon cathode (specific surface area $66 cm^2 cm^{-3}$) and (d) a $5 \times 5 \times 1.2$ cm carbon felt cathode. Data taken from D. Pletcher, I. Whyte, F. C. Walsh and P. J. Millington, *J. Appl. Electrochem.*, 1991, **21**, 659.

compared to the flat plate. With the carbon felt the current scales by a factor > 1000, reaching a current density of $1 A cm^{-2}$ (based on the geometric area of the face of the cathode). These plots demonstrate the possibility of a much more rapid electrolysis or, if the objective is to remove the metal ion in a single pass, to a much shorter cell. For example, the data of Figure 9.7 were used to calculate that 99% of the Cu^{2+} could be removed in a single pass from this acid solution if the cathode was fabricated from 100 ppi reticulated vitreous carbon and it had a height of 40 cm. Figure 9.8(a) shows a sketch of a cell built to test this result; it has eight reticulated vitreous carbon blocks, each 5 cm high, and there were sample points between each. Figure 9.8(b) shows a plot of Cu^{2+} concentration as a function of distance through the cell and it can be seen that the performance of the cell corresponds closely to the theoretical prediction.

It is important to stress that the current will not always scale with the dimensions of the three-dimensional electrode in the direction parallel to charge flow (in the cell of Figure 9.8a, the thickness of the reticulated vitreous carbon blocks). The current flowing in the solution within the electrode structure will lead to an *IR* drop; the driving force for electron transfer will be highest at the front of the three-dimensional electrode

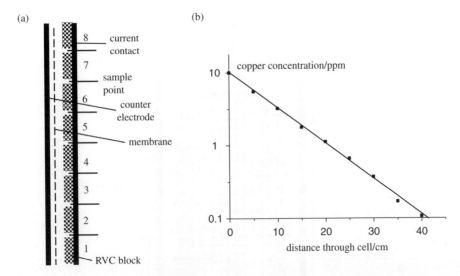

Figure 9.8 (*a*) Cell with 100 ppi reticulated vitreous carbon cathode for the removal of copper from a solution containing 10 ppm Cu^{2+} + 0.5 M Na_2SO_4, pH 2; (*b*) plot of copper concentration as a function of distance through the cathode. Data taken from D. Pletcher, I. Whyte, F. C. Walsh and P. J. Millington, *J. Appl. Electrochem.*, 1993, **23**, 82.

and lowest at the interface with the current contact. As a result, if the electrode is made too thick, not all the electrode will operate under mass transport control and, in the limit, there may be a region where no reaction occurs. The useful thickness will depend on the conductivity of the solution and the magnitude of the local current density, *i.e.* the mass transfer coefficient and, more importantly, the concentration of the reactant. Three-dimensional electrodes will be most effective with low reactant concentrations. They are therefore well matched to the needs of effluent treatment.

From the data in Figure 9.7 it is apparent that a cathode fabricated from carbon felt could perform even better than that based on reticulated vitreous carbon. This was recognized and the challenge is to design a cell where the carbon felt is used to its full advantage without a penalty associated with poor solution flow and to include the cell in robust, inexpensive and user-friendly units for effluent treatment. These concepts led to the development of the Porocell™, later to become the Renocell™. Figure 9.9(a) shows the cell design; several sizes are available. The cells are constructed within a polymer pipe. The cathode is a cylinder of carbon felt (~ 1 cm thick) supported on a polymer mesh and the concentric metal gauze anode (often Pt/Ti) is outside; it is possible to include a separator but this is avoided if possible. The effluent is pumped

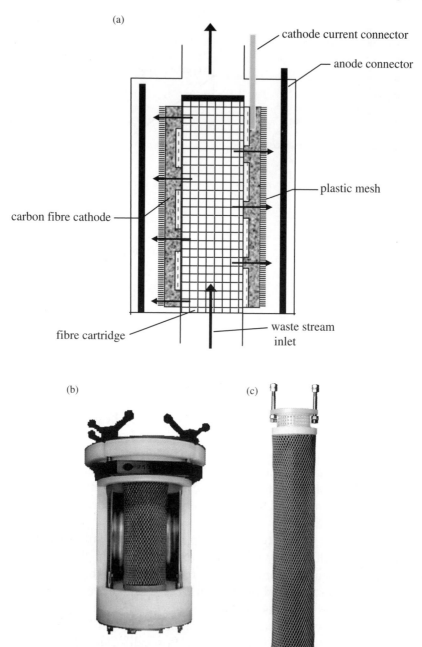

(a)

cathode current connector

anode connector

plastic mesh

carbon fibre cathode

fibre cartridge

waste stream inlet

(b)

(c)

Figure 9.9 (*a*) Schematic diagram of the Renocell™ used for the removal of heavy and transition metal ions from effluent; (*b*) cutaway of a M250 Renocell; (*c*) M500 cathode cartridge. Artwork courtesy of Renovare International Inc.

up the centre of the carbon felt cylinder and flows outwards through the cathode, the relatively short electrolyte path through the cathode ensuring a uniform electrolyte flow. Effluent is therefore treated in batches with recycle through the cell as a high conversion cannot be achieved in a single pass. The metal from the effluent is deposited within the carbon felt but a surprisingly large quantity can be accommodated without blockage of the structure. The unique features of the system are that the cathode is constructed as an easily replaceable cartridge and the cell is

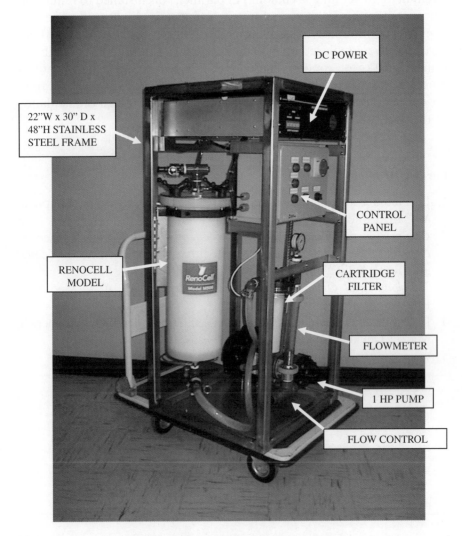

DC POWER

22"W x 30" D x 48"H STAINLESS STEEL FRAME

CONTROL PANEL

RENOCELL MODEL

CARTRIDGE FILTER

FLOWMETER

1 HP PUMP

FLOW CONTROL

Figure 9.10 A Renocell™ Model MX0501 system for the removal of heavy and transition metal ions from waste streams. Photograph courtesy of Renovare International Inc.

connected to the reservoir and pump circuit *via* quick release couplings, making it easy to remove the filled cathode cartridges and to install replacements. Figure 9.9(b) and (c) shows a cutaway of a Renocell™ cell and a cathode cartridge. Figure 9.10 shows a complete unit.

An early application was the recovery of copper in a distillery for the production of Scotch whiskey. The distillation is carried out in copper pot stills and, in this distillery, this led to $\sim 7000 \, L \, h^{-1}$ of a waste stream containing 10–30 ppm Cu^{2+}. A unit, in fact containing five cells (each 18 cm in diameter and 50 cm long) in parallel, was installed to reduce the Cu^{2+} content to < 1 ppm before the effluent is discharged. The unit operated continuously, removing $\sim 2 \, kg$ copper per day. The cathode cartridges were replaced on a 30 day cycle when they had each gathered some 10–20 kg of copper. The copper is recovered in a pure state by burning away the carbon felt in a furnace. Over a relatively short time, the copper metal recovered was sufficient to cover the cost of instillation of the unit.

Further Reading

1. D. Pletcher and F. C. Walsh, *Industrial Electrochemistry*, Chapman & Hall, London, 1990.
2. *Comprehensive Treatise of Electrochemistry, Volume 2, Electrochemical Processing*, ed. J. O'M. Bockris, B. E. Conway, E. Yeager and R. E. White, Plenum, New York, 1981.
3. *Environmental Orientated Electrochemistry*, ed. C. A. C. Sequeira, Elsevier, Amsterdam, 1994.
4. H. Pütter, in *Organic Electrochemistry*, ed. H. Lund and O. Hammerich, Marcel Dekker, New York, 2001, p. 1259.
5. www.renovare.com, Jan. 2009.

CHAPTER 10
Problems and Solutions

This chapter consists of 26 problems and complete answers. The problems are designed to test the reader's grasp of the fundamental concepts of electrochemistry and ability to interpret data from electrochemical experiments. As far as possible, the problems are based on real experimental data from the laboratory. Real benefit will result only by trying the problems before looking at the solutions!

Throughout, to simplify the arithmetic, I suggest that you assume that $2.3RT/F = 60\,mV$ at $298\,K$ and that the Faraday constant is $96500\,C\,mol^{-1}$. Also, take the equilibrium potentials for the H^+/H_2 and O_2/H_2O as 0 and $+1230\,mV$ vs RHE respectively.

10.1 PROBLEMS

1. The formal potentials of the $Fe(III)/Fe(II)$ couple in three acids at pH 0 were measured as:

HCl	$+524\,mV$ vs a Hg/Hg_2Cl_2 in saturated KCl (SCE) electrode
H_2SO_4	$+39\,mV$ vs a Hg/Hg_2SO_4 in saturated K_2SO_4 electrode
H_3PO_4	$+438\,mV$ vs SHE

The potentials of the saturated calomel and saturated mercurous sulfate reference electrodes are $+246$ and $+640\,mV$ vs SHE, respectively.

A First Course in Electrode Processes, 2nd Edition
By Derek Pletcher
© Derek Pletcher 2009
Published by the Royal Society of Chemistry, www.rsc.org

(a) How would you measure these formal potentials?

(b) In which medium is Fe(III) the strongest oxidizing agent?

2. Compare the free energies for the formation of 1 mole of copper in undivided, electrowinning cells where the anode reaction is O_2 evolution and the electrolytes are shown in the table (EDTA is the ligand ethylenediaminetetraacetate):

1 M Cu(II) in 2 M sulfate, pH 0	E_e° for Cu(II)/Cu $= +142$ mV vs SHE
1 M Cu(II)EDTA in buffer, pH 5	E_e° for Cu(II)/Cu $= -170$ mV vs SHE
1 M Cu(I) in 5 M chloride, pH 1	E_e° for Cu(I)/Cu $= -84$ mV vs SHE

3. Two primary batteries are based on the cell reactions:

$$2MnO_2 + Zn + 2NH_4Cl \longrightarrow Mn_2O_3 + Zn(NH_3)_2Cl_2 + H_2O$$

and:

$$2Li + 2SO_2 \longrightarrow Li_2S_2O_4$$

that have free energies of $-229\,kJ\,mol^{-1}$ of Zn and $-294\,kJ\,mol^{-1}$ of Li, respectively. Typical batteries have a capacity of 2 Ah (ampere hours). Estimate the battery voltages and the weights of the two metals needed to deliver this capacity. Do your calculations reveal any advantages of the lithium battery? The atomic weights of zinc and lithium are 65 and 7, respectively.

4. The voltammogram below is recorded at a gold RDE (area 0.2 cm², rotation rate 3600 rpm) for a solution containing 100 mM Ce(IV) + 50 mM Ce(III) in 3 M CH_3SO_3H.

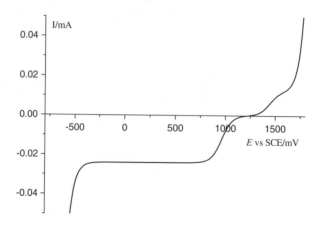

Identify all the electrode reactions taking place and indicate the rate-determining steps in each potential range. Assuming that the plateau currents are proportional to the square root of the rotation rate of the disc, estimate the mass transfer coefficient from the experimental data and compare the value with that from the theoretical equation.

5. Using the same electrode and solution as in question 4 above, the following steady state data were obtained:

E vs SCE/mV	1290	1270	1250	1160	1140	1120	1100	1080	1050
I/mA	+0.91	+0.62	+0.42	–0.39	–0.57	–0.84	–1.23	–1.81	–3.20

Estimate values for the equilibrium potential of the solution and the formal potential, exchange current density, standard rate constant and transfer coefficient for the $Ce(IV)/Ce(III)$ couple.

6. The redox couple $Eu(III)/Eu(II)$ has a formal potential of $-596\,mV$ *vs* SCE and a standard rate constant of $2 \times 10^{-3}\,cm\,s^{-1}$ in 1 M $NaClO_4$, pH 2. Sketch the *j vs E* characteristic (including scales on the current density and potential axes) to be expected at an amalgamated disc electrode rotating at 1600 rpm for a solution containing 10 mM $Eu(III)$ and 1 mM $Eu(II)$. Assume that $\alpha = 0.5$, $D = 6 \times 10^{-6}\,cm^2\,s^{-1}$ and the kinematic viscosity is $10^{-2}\,cm^2\,s^{-1}$. Locate the equilibrium potential and discuss the way in which the current density will vary with rotation rate of the RDE.

7. The reaction $Fe(III) + e^- \rightleftharpoons Fe(II)$ has a formal potential of $+679\,mV$ *vs* SHE in 1 M H_2SO_4, a standard rate constant of $10^{-6}\,cm\,s^{-1}$ and a transfer coefficient of 0.5. For a solution containing 10 mM $Fe(III)$ and 1 mM $Fe(II)$ calculate the equilibrium potential *vs* SCE and the exchange current density. Then (a) use the Tafel equation to calculate current densities at $+505$, 613 and 853 mV *vs* SCE; (b) compare the values with those obtained using the Butler–Volmer equation and the value of the mass transfer controlled current (assume $k_m = 2 \times 10^{-3}\,cm\,s^{-1}$) and then comment on the differences; and (c) sketch a voltammogram for the solution (including scales on the current density and potential axes). Why does the voltammogram differ from that in the previous question?

8. Assuming that $D = 6 \times 10^{-6}\,cm^2\,s^{-1}$ and the kinematic viscosity is $10^{-2}\,cm^2\,s^{-1}$, estimate values for the mass transfer coefficient for:

(a) a rotating disc electrode, $\omega = 900$ and 3600 rpm, (b) a stationary planar electrode, 1 ms and 10 s after the imposition of a potential step and (c) microdisc electrodes with radii 1 and 25 μm. How do these calculations assist the selection of a technique for the study of the kinetics of electron transfer?

9. In non-complexing aqueous solutions, the standard potentials of the Cu(II)/Cu(I) and Cu(I)/Cu couples are $+167$ and $+522$ mV *vs* SHE, respectively. In acetonitrile, the values for the Cu(II)/Cu(I) and Cu(I)/Cu couples are $+1194$ and -380 mV, respectively. Calculate the equilibrium constants for the reaction:

$$Cu(II) + Cu \rightleftharpoons 2Cu(I)$$

in the two media. What do the values say about the chemistry of Cu(I)? Suggest a reason for the difference observed.

10. Calculate the potential of the cathode during the electroplating of copper at 20 mA cm^{-2} from solutions containing 1 M of the copper species:
    (a) using an acid sulfate bath – the formal potential is $+340$ mV *vs* SHE, the exchange current density is 0.1 mA cm^{-2} and the Tafel slope $(120\,\text{mV})^{-1}$;
    (b) using a cyanide bath containing 5 M cyanide and where the principal copper species is Cu(CN)$_4^{3-}$. The stability constant for this complex is 10^{25}. The formal potential of the Cu(I)/Cu couple in non-complexing media is $+522$ mV *vs* SHE. In the cyanide bath, the exchange current density is 0.01 mA cm^{-2} and the Tafel slope $(120\,\text{mV})^{-1}$

11. The formal potential for the Cu(I)/Cu couple as a function of ammonia concentration in aqueous solutions were measured:

[NH$_3$]/M	0.01	0.03	0.1	0.3	1.0	3.0
E_e° *vs* SCE/mV	-260	-315	-376	-452	-502	-567

Determine the major Cu(I) complex in solution and calculate a value for the stability constant for this complex. The formal potential for the Cu(I)/Cu couple in non-complexing media is $+276$ mV *vs* SCE.

12. A parallel plate, flow cell with a membrane between anolyte and catholyte is set up for the synthesis of a fine chemical at the cathode. The cell has (a) a graphite cathode plate, thickness 1 cm, (b) two

electrolyte chambers where the membrane/electrode separation is 0.5 cm, (c) a proton-conducting membrane, 0.01 cm thick and (d) a Ti plate anode (0.2 cm thick) with a coating of IrO_2-based catalyst (5 μm thick). The anolyte is 1 M H_2SO_4 and the catholyte an acetate buffer pH 4.8 in ethanol/water. Calculate the voltage drop through the cell when the current density is 0.2 A cm^{-2} and suggest tactics for decreasing it. At the operating temperature of 318 K, the following conductivities are available:

	$\kappa/$S cm^{-1}
Graphite	2.1×10^2
Titanium	1.8×10^4
Iridium dioxide	4.6×10^{-3}
1 M H_2SO_4	0.74
Ethanol/water + buffer	0.05
Membrane	0.07

13. A chromic acid etching solution is regenerated by the anodic oxidation of Cr(III) in the sulfuric acid medium using a membrane cell and a current density of 150 mA cm^{-2}. Using the following data estimate the cell voltage and energy efficiency for the cell:
   (a) the electrode/membrane gaps are 0.4 cm and the electrolyte conductivities 0.42 S cm^{-1},
   (b) the membrane resistance is quoted as 1.8 ohm cm^2,
   (c) the free energy for the cell reaction:

$$2Cr^{3+} + 7H_2O \longrightarrow Cr_2O_7{}^{2-} + 3H_2 + 8H^+$$

   is 852 kJ mol^{-1} of dichromate,
   (d) the overpotentials at the anode and cathode were estimated from $j - E$ data as 340 and -210 mV respectively.

The plant has a bipolar cell stack containing 25, 1 m^2 electrodes. Estimate the annual production of $H_2Cr_2O_7$ (molecular weight 218).

14. The interfacial tension, γ, of the interface between mercury and 0.1 M aqueous NaI was measured as a function of potential:

ΔE/mV	+ 100	+ 50	0	-100	-200	-300	-400
$\gamma/$N cm^{-1}	3.25	3.45	3.60	3.78	3.88	3.94	3.91
ΔE/mV	-500	-600	-700	-800	-900	-1000	-1200
$\gamma/$N cm^{-1}	3.83	3.73	3.62	3.50	3.35	3.18	2.70

Here ΔE is defined as the shift in potential from the potential of zero charge for mercury in 0.1 M NaF. Estimate (a) the potential of zero charge for the 0.1 M NaI solution and (b) the charge on the interface at 0 and $-800\,mV$. Comment on the data.

15. Voltammograms were recorded at a rotating vitreous carbon disc electrode (area $0.12\,cm^2$) in a solution of 2 mM nitro-benzene $+ 0.2$ M Bu_4NBF_4 in acetonitrile. A well-formed reduction wave was observed at $E_{1/2} = -1.10$ V *vs* SCE and the plateau current was found to vary with rotation rate:

ω/rpm	400	900	1600	2500	3600
$-I_L/\mu A$	130	195	260	325	395

The following data were taken from the voltammogram at 400 rpm:

$-E$ *vs* SCE/mV	1060	1080	1100	1120	1140	1160	1180
$-I/\mu A$	12	23	40	67	83	101	120

After the addition of 50 mM ethanoic acid, the first reduction wave had shifted positive to -850 mV *vs* SCE and the variation in the limiting current with rotation was:

ω/rpm	400	900	1600	2500	3600
$-I_L/\mu A$	510	790	1040	1310	1595

Interpret this data as quantitatively as possible. The kinematic viscosity of the electrolyte is $3.7 \times 10^{-3}\,cm^2\,s^{-1}$.

16. Voltammograms for four rotation rates are presented in the figure for a Au electrode in oxygen saturated, 0.5 M $HClO_4$.

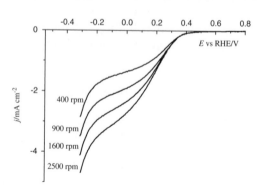

Current density as a function of potential and rotation rate, taken from these experimental data, are given in the table below:

	j/mA cm^{-2} at rotation rates of			
Potential/mV	**400 rpm**	**900 rpm**	**1600 rpm**	**2500 rpm**
+400	−0.090	−0.090	−0.090	−0.090
+350	−0.182	−0.199	−0.206	−0.212
+300	−0.383	−0.434	−0.467	−0.493
+200	−0.520	−0.740	−0.930	−1.090

Comment on the data. Estimate the kinetically controlled current density at each potential and investigate whether a plot of log(kinetic current) *vs* potential is linear.

17. The figures show voltammograms at a rotating vitreous carbon disc electrode (area 0.12 cm^2) in a deoxygenated solution of 2 mM Cu(II) in 1.5 M NaCl, pH 3. (a) Shows the influence of rotation rate on the voltammogram for the first reduction process. (b) Shows the influence of the direction of the potential scan, $\omega = 900$ rpm. (c) Shows the voltammogram with an extended negative potential limit, $\omega = 900$ rpm. Explain all the features of the data.

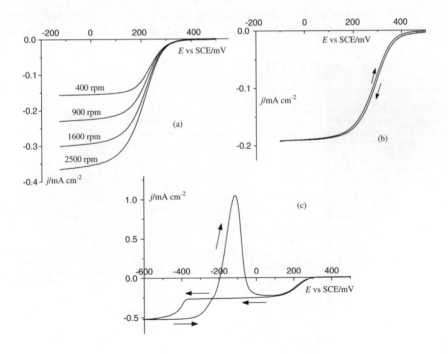

18. The figure reports a cyclic voltammogram (potential scan rate $100\,mV\,s^{-1}$) recorded for a stationary vitreous carbon disc (area $0.12\,cm^2$) in the deoxygenated solution of $2\,mM$ Cu(II) in $1.5\,M$ NaCl, pH 3. Explain the change in shape of the voltammogram to that in the previous question. What features allow confirmation of the conclusions you drew from the data at the rotating disc electrode?

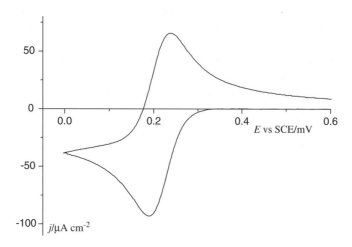

19. A vitreous carbon disc/gold ring electrode was calibrated using a solution of $5\,mM$ $Fe(CN)_6^{4-} + 0.5\,M$ KNO_3; the figure show a voltammogram at the disc at 600 rpm with the response at the ring held at $-200\,mV$ *vs* SCE.

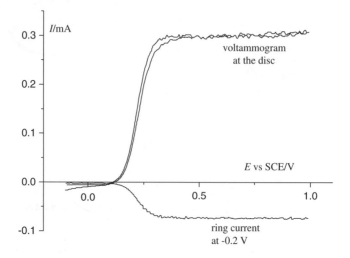

The RRDE was then used to study O_2 reduction at vitreous carbon. The figure shows, for O_2 saturated 0.1 M KOH, the voltammogram at the disc and ring response when the potential is held at $+800$ mV *vs* SCE (rotation rate 600 rpm).

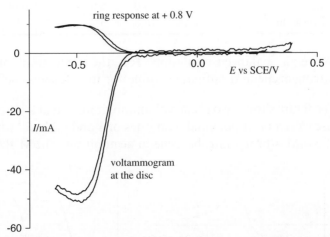

What conclusion can be drawn about O_2 reduction at vitreous carbon?

20. The corrosion of copper in O_2 saturated, aqueous 1 M NaCl, pH 1, was investigated using a rotating Cu disc (area $0.18\,cm^2$)/Au ring electrode with a collection efficiency equal to 0.34. With the copper disc electrode on open circuit, the RRDE was rotated at 900 rpm for 30 min (100 cm^3 of the electrolyte); after this period, a sample of the solution was taken and analysed by atomic absorption spectroscopy and shown to contain $0.93\,\mu g\,cm^{-3}$. During the 30 min period, the average ring current (potential $+450$ mV *vs* SCE) was $+26\,\mu A$. Two further observations were made: (a) in the oxygen saturated solution, the ring current is independent of rotation rate of the RRDE; (b) on purging the solution with N_2, the ring current was zero. Write down the reactions occurring at the disc and ring. What does the data tell us about the corrosion of copper under these conditions?

21. 2-Fluoronitrobenzene (145 mg; molecular weight 141) was dissolved in 50/50 ethanol/water containing 1 M sulfuric acid and reduced at a mercury cathode (area $10\,cm^2$) at -800 mV. The following data were noted:

Charge passed/C	100	200	300	400	500
Current/mA	-98	-81	-59	-39	-22

A potential step experiment from −300 to −800 mV *vs* SCE at a Hg drop electrode in a solution of 1.2 mM 2-fluoronitrobenzene in the same medium led to a falling transient:

t/ms	1.0	4.0	9.0	36.0	144	1000
$-j$ /mA cm^{-2}	33.1	16.3	10.8	5.7	2.9	1.1

Write a balanced equation for the reduction of 2-fluoro-nitrobenzene and estimate a value for its diffusion coefficient.

22. The figure shows two cyclic voltammograms run at a vitreous carbon disc electrode at potential scan rates of 1 and 0.1 V s^{-1} for a solution of 3 mM 4-bromonitrobenzene in acetonitrile + 0.2 M Bu$_4$NBF$_4$.

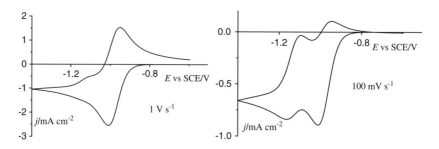

Under the same conditions, nitrobenzene gives a reversible 1e$^-$ reduction process centred around −1.13 V *vs* SCE. Suggest a mechanism for the reduction of 4-bromonitrobenzene and additional experiments that could be used to confirm the mechanism.

23. The figure shows two cyclic voltammograms (100 mV s^{-1} at a Pt electrode for compound A, a drug intermediate, in acetonitrile + 0.2 M Bu$_4$NBF$_4$.

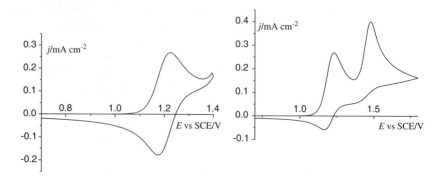

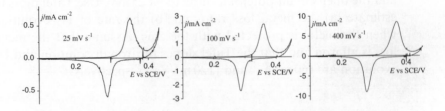

A

During a controlled potential electrolysis at +1600 mV of a solution containing 5.3 mmol of A, the following data were obtained:

Q/C	100	198	412	604	788
I/mA	81	71	52	35	20

It is postulated that at +1600 mV, the reaction is:

$$A + 2H_2O - 2e^- \longrightarrow$$

+ + CH_3OH + $2H^+$

Are the experimental data compatible with this proposal? Suggest additional experiments to confirm a mechanism.

24. The figure shows a set of cyclic voltammograms for a Ni electrode in 1 M KOH.

The voltammograms are unchanged when run repetitively without cleaning the electrode. Recognizing that when nickel is placed into an alkaline solution, a spontaneous and rapid reaction occurs to form a surface layer of $Ni(OH)_2$, suggest an interpretation of the voltammograms.

25. The voltammogram (potential scan rate $50\,mV\,s^{-1}$) for a nickel disc in an aqueous solution of phosphoric acid, pH 2, at 353 K is shown in the figure. If a second scan is carried out without repolishing the

Ni surface, the peak is much diminished and on repetitive cycling the anodic current drops to a low value. Suggest an interpretation of this electrochemistry.

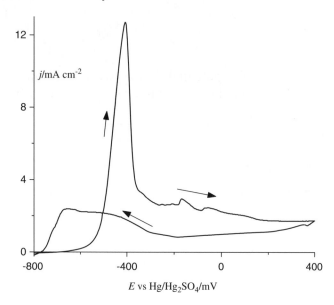

26. A carbon steel structure in an oxygen free, aqueous buffer, pH 4, was found to have a (open circuit) corrosion potential of $-510\,mV$ *vs* SCE and a metal loss rate of $0.3\,mg\,cm^{-2}\,year^{-1}$. To decrease the corrosion rate, a sacrificial zinc anode is attached to the structure and the open circuit potential shifts to $-825\,mV$. Use Tafel plots to estimate (a) the metal loss rate and (b) the rate of H_2 evolution when the metal is protected with the zinc. Assume that the metal loss is all iron and that the Tafel slopes for iron dissolution and H_2 evolution are $(60\,mV)^{-1}$ and $(120\,mV)^{-1}$, respectively.

10.2 SOLUTIONS

1. (a) Prepare solutions in each of the acids, pH 0, with equal concentrations of Fe(III) and Fe(II) (say, 5 mM). Place each solution into a cell with an inert working electrode (*e.g.* a vitreous carbon disc) and the reference electrode. Remove O_2 from the solution by passing a fast stream of fine nitrogen bubbles through the solution. Measure the potential difference between the two electrodes with a high impedance digital voltmeter.

   (b) The formal potentials must be converted into values *versus* a single reference electrode:

	E_e^0 vs SHE/mV
HCl	$524 + 246 = 770$
H_2SO_4	$39 + 640 = 679$
H_3PO_4	438

If in doubt, draw a potential diagram for each medium, *e.g.* for HCl:

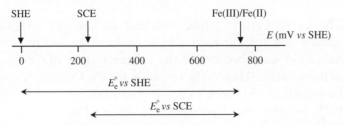

Fe(III) is a stronger oxidizing agent in HCl since this is the medium where the formal potential is most positive (if this is not obvious to you, consider the free energies of the reactions, $Fe(III) + \frac{1}{2}H_2 \leftrightarrow Fe(II) + H^+$ in each of the three solutions). The differences in the formal potentials arise from differences in speciation with the different anions present; either complexing the Fe(III) more strongly or the Fe(II) more weakly will cause the formal potential to shift negative.

2. The equilibrium potential for the oxygen anode shifts negative by 60 mV per pH unit (as calculated from the Nernst equation). Hence, the equilibrium cell voltage may be calculated, then the free energies using $\Delta G = -nFE_{cell}^0$ and noting the change in n.

Medium	E_{cell}^0/mV	n	$\Delta G/\text{kJ mol}^{-1}$ of Cu
Sulfate	$142 - 1230 = -1088$	2	210
EDTA	$-170 - 930 = -1100$	2	212
Chloride	$-84 - 1170 = -1254$	1	121

Although the equilibrium cell voltages are very similar, the free energy for the formation of 1 mole of copper is much less in the chloride medium where the copper is present as Cu(I). Provided there are no kinetic complications, a process based on Cu(I) and a chloride medium would consume much less energy.

3. The cell voltages are again calculated using the equation, $\Delta G = -nFE^0_{cell}$ while the weight of metal required is a Faraday's law calculation.

Battery	Cell voltage/V	Weight of metal/g
Zn/MnO$_2$	$\dfrac{299 \times 10^3}{2 \times 96500} = 1.55$	$\dfrac{2 \times 3.6 \times 10^3}{2 \times 96500} \times 65 = 2.44$
Li/SO$_2$	$\dfrac{294 \times 10^3}{96500} = 3.05$	$\dfrac{2 \times 3.6 \times 10^3}{96500} \times 7 = 0.52$

It can be seen that the lithium battery has a higher voltage and that the capacity is achieved with a much lower weight of reactant.

4. The Ce(IV)/Ce(III) couple will lead to both an oxidation and a reduction wave, the reduction wave having twice the height of the oxidation wave because of the concentrations of Ce(III) and Ce(IV) in the solution. Hence, the response for the Ce(IV)/Ce(III) couple can be identified. At more extreme potentials, current is observed for both oxygen and hydrogen evolution. The oxidation wave for Ce(III) does not give a clear limiting current because the potential range overlaps with that for O$_2$ evolution.

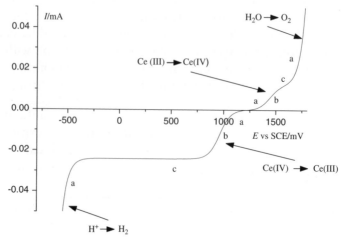

The rate-determining steps are as labelled: a = electron-transfer control, b = mixed control and c = mass-transport control. The mass transfer coefficient can best be estimated from the experimental current in the good limiting current region for the reduction of Ce(IV) and Equation (1.61), *i.e.*:

$$k_m = \frac{I_L}{nFAc} = \frac{0.024}{96500 \times 0.2 \times 10^{-4}} = 1.25 \times 10^{-2} \, \text{cm s}^{-1}$$

The theoretical value is calculated from Equation (7.8), *i.e.*:

$$k_m = \frac{D^{2/3}\omega^{1/2}}{1.61\nu^{1/6}} = \frac{\left(6 \times 10^{-6}\right)^{2/3}\left(\dfrac{2\pi \times 3600}{60}\right)^{1/2}}{1.61(10^{-2})^{1/6}} = 8.5 \times 10^{-3} \, \text{cm s}^{-1}$$

This is reasonable agreement in view of the assumptions made (*e.g.* the value of the diffusion coefficient).

5. Plotting the current *vs* potential data as a Tafel plot:

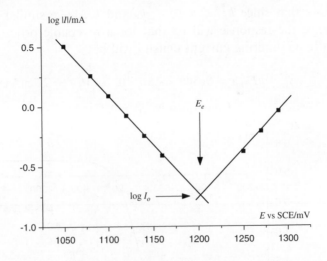

It can be seen that both the oxidation and reduction data give linear Tafel plots. Reading from the plot:

$E_e \approx 1200$ mV and log $I_o \approx -0.75$ and the slopes are close to $(120 \, \text{mV})^{-1}$, $\alpha = 0.5$.

The formal potential is estimated from the equilibrium potential using the Nernst equation, (1.29); $E_e^0 = 1180$ mV.

The exchange current density, $j_o = 5 \times$ antilog $-0.7 = 1\,\text{mA}\,\text{cm}^{-2}$. The standard rate constant,

$$k_s = \frac{10^{-3}}{96500(10^{-4} \times 5 \times 10^{-5})^{1/2}} = 1.5 \times 10^{-4}\,\text{cm}\,\text{s}^{-1}$$

Note: In this system, kinetically controlled currents can be measured only over a rather narrow range because of background currents and mass transport limitations. This reduces the reliability of the values for the kinetic constants calculated.

6. Using the Nernst equation, (1.29), $E_e = -596 + 60 = 536\,\text{mV}$. The mass transfer coefficient is again given by Equation (7.8):

$$k_m = \frac{D^{2/3}\omega^{1/2}}{1.61\nu^{1/6}} = \frac{(6 \times 10^{-6})^{2/3}\left(\dfrac{2\pi \times 400}{60}\right)^{1/2}}{1.61(10^{-2})^{1/6}} = 2.8 \times 10^{-3}\,\text{cm}\,\text{s}^{-1}$$

Note that since $k_s = 2 \times 10^{-3}$, k_s and k_m have similar values and hence the response will be that for a reversible process. Also the cathodic limiting current density will be:

$$j_L^C = nF k_m c = 96500 \times 2.8 \times 10^{-3} \times 10^{-5} = 2.7\,\text{mA}\,\text{cm}^{-2}.$$

Consequently, $j_L^A = 0.27\,\text{mA}\,\text{cm}^{-2}$.

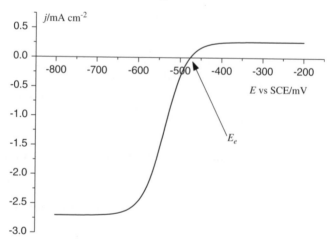

Since the reaction is reversible, the current will be proportional to the square root of rotation rate of the disc electrode at *all* potentials.

7. Using the Nernst equation, (1.29), $E_e = 679 + 60 = +739$ mV *vs* SHE $= +493$ mV *vs* SCE. The potentials of 505, 613 and 853 mV *vs* SCE, therefore correspond to overpotentials of 12, 120 and 360 mV. The exchange current density is calculated from (1.45):

$$j_o = 96500 \times 10^{-6} \times (10^{-5} \times 10^{-6})^{1/2} = 0.36 \, \mu\text{A cm}^{-2}$$

Also for the oxidation, $j_L = nFk_m c = 200 \, \mu\text{A cm}^{-2}$

$\eta/$ mV	$j = j_o \exp\dfrac{\alpha nF\eta}{RT}$ / $\mu\text{A cm}^{-2}$	$j = j_o\left(\exp\dfrac{\alpha nF\eta}{RT} - \exp -\dfrac{(1-\alpha)nF\eta}{RT}\right)$ / $\mu\text{A cm}^{-2}$
12	0.45	0.17
120	3.62	3.61
360	365	365

Conclusions: at $+505$ mV, both reduction as well as oxidation partial current densities must be considered and hence the Butler–Volmer equation should be used. By 613 mV, the anodic current density dominates and the Tafel equation is sufficient. At 853 mV, the oxidation of Fe(II) is mass transport controlled.

The voltammogram differs from the Eu(III)/Eu(II) case because of the low value of the standard rate constant. The Fe(III)/Fe(II) couple is irreversible and the oxidation and reduction waves will be separated on the potential axis.

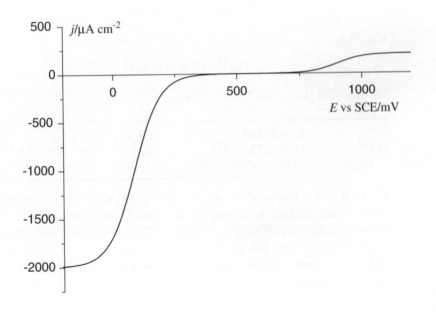

8.   (a) At the RDE:

$$k_m = \frac{D^{2/3}\omega^{1/2}}{1.61\nu^{1/6}} = \frac{(6 \times 10^{-6})^{2/3} \left(\dfrac{2\pi \times 900}{60}\right)^{1/2}}{1.61(10^{-2})^{1/6}} = 4.2 \times 10^{-3}\,\text{cm}\,\text{s}^{-1}$$

with a rotation rate of 900 rpm. With $\omega = 3600$ rpm, $k_m = 8.4 \times 10^{-3}$ cm s^{-1}.

(b) Under conditions of planar diffusion, $k_m = (D/\pi t)^{1/2}$ by comparing the Cottrell equation, (7.36), with $j_L = nFk_m c$. At 1 ms and 10 s, k_m is 4.4×10^{-2} and 4.4×10^{-4} cm s^{-1}, respectively.

(c) The steady state mass transfer coefficient to a microdisc electrode is given by $k_m = 4D/\pi t^{1/2}$, Equation (7.74). Hence for 1 μm and 25 μm radius discs, k_m is 7.6×10^{-2} and 3×10^{-3} cm s^{-1}, respectively.

A small microdisc or a short timescale experiment at a planar electrode gives a much higher rate of mass transport than a RDE and hence are more suitable for the study of fast electron transfer.

9. Subtracting the reaction:

$$Cu(I) + e^- \rightleftharpoons Cu \tag{1}$$

from:

$$Cu(II) + e^- \rightleftharpoons Cu(I) \tag{2}$$

leads to:

$$Cu(II) + Cu \rightleftharpoons 2Cu(I)$$

Noting that $\Delta G = -nFE = -2.3RT\log K_{eq}$:

	$E_2 - E_1$/mV	Log K_{eq}	K_{eq}
H_2O	$167 - 522 = -355$	$-355/60$	1.2×10^{-6}
CH_3CN	$1194 - (-380) = 1574$	$1574/60$	1.7×10^{26}

There is a very large difference; in water Cu(I) will disproportionate while in acetonitrile it is very stable. This arises because Cu(I) has a strong affinity for nitrogen ligands. Hence, acetonitrile solvates Cu(I) strongly and greatly increases its stability, making it more difficult to form either Cu(II) or Cu and this is reflected in the formal potentials.

10. (a) Sulfate medium. The overpotential required is calculated from the Tafel equation, (1.46); $\eta = -276\,\text{mV}$ and $E = 340 - 276 = 64\,\text{mV}$ *vs* SHE.

 (b) In this solution, the equilibrium:

$$Cu(I) + 4CN^- \rightleftharpoons Cu(CN)_4^{3-}$$

has to be taken into account.

$$K_{eq} = \frac{[Cu(CN)_4^{3-}]}{[Cu(I)][CN^-]^4} = 10^{25}$$

In the bath $[Cu(CN)_4^{3-}] = 1\,\text{M}$ and $[CN^-] = 5\,\text{M}$. Hence $[Cu(I)] = 1.6 \times 10^{-28}\,\text{M}$ and using the Nernst equation: $E_e = 522 + 60\,\log[Cu(I)] = 522 - 60 \times 27.8 = -1146\,\text{mV}$. Using the Tafel equation, $\eta = -396$ and $E = -1146 - 396 = -1542\,\text{mV}$ *vs* SHE.

11. The complex is formed through the equilibrium:

$$Cu(I) + yNH_3 \rightleftharpoons Cu(NH_3)_y{}^+$$

with an equilibrium constant:

$$K_{eq} = \frac{[Cu(NH_3)_y{}^+]}{[Cu(I)][NH_3]^y}$$

The Nernst equation for the $Cu(I)/Cu$ couple may be written in terms of either uncomplexed or complexed $Cu(I)$:

$$E_e = 276 + 60\,\log[Cu(I)] = (E_e^\circ)_{complex} + 60\,\log[Cu(NH_3)_y{}^+]$$

Using the equation for the equilibrium potential and then simplifying leads to:

$$276 - (E_e^\circ)_{complex} = 60\,\log K_{eq} + y60\,\log[NH_3]$$

Plotting $(E_e^\circ)_{complex}$ *vs* $\log[NH_3]$ therefore allows determination of y from the slope and K_{eq} from the intercept.

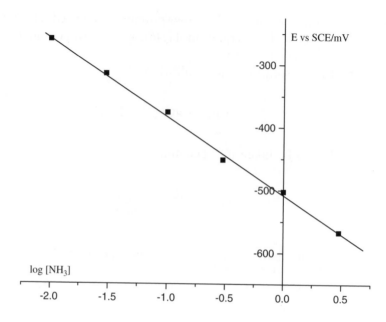

$$y = \frac{120}{60} = 2 \text{ and } \log K_{eq} = \frac{276 - (-510)}{60} \text{ so } K_{eq} = 1.3 \times 10^{13}$$

12. Using the expression $R = S/\kappa A$, the following table may be drawn up for $j = 0.2\,\text{A cm}^{-2}$:

Component	$R/\text{ohm cm}^2$	jR/mV
Graphite cathode	1/210	1
Titanium anode	$0.2/1.8 \times 10^4$	0.002
Anode coating	$5 \times 10^{-4}/4.6 \times 10^{-3}$	22
Catholyte	0.5/0.05	2000
Anolyte	0.5/0.74	135
Membrane	0.01/0.07	29

Clearly, the dominant jR drop is that through the catholyte. This voltage drop could be decreased by increasing the catholyte conductivity (by, for example, reducing the ethanol content, increasing the temperature) and decreasing the cathode membrane gap. The voltage drops through the electrodes are negligible and small through the membrane and anode coating.

13. The contributions to the cell voltage can be tabulated:

Thermodynamic potential	$\dfrac{852 \times 10^3}{6 \times 96500} =$	-1.472 V
Overpotentials	$-(340 + 210) =$	-0.550 V
Electrolyte voltage drop	$-2 \times \dfrac{0.4}{0.42} \times 0.15 =$	-0.285 V
Membrane voltage drop	$-1.8 \times 0.15 =$	-0.270 V
Cell voltage		-2.577 V

Note: (a) the formation of one dichromate requires the oxidation of two Cr(III) in a $6e^-$ reaction; (b) both overpotentials increase the cell voltage. Hence, the cell voltage is -2.577 V and the energy efficiency $= (1.472/2.577) \times 100 = 57\%$. Only 57% of the energy input is necessary thermodynamically and 43% is inefficiency that results in heat.

$$\text{Charge passed/year} = 25 \times 10^4 \times 0.15 \times 3600 \times 24 \times 365$$
$$= 1.18 \times 10^{12} \text{ C}$$

$$\text{Annual production} = \frac{1.18 \times 10^{12}}{6 \times 96500} 218 = 4.45 \times 10^8 \text{g} = 445 \text{ tonnes}$$

14. Plotting the interfacial tension *vs* the potential:

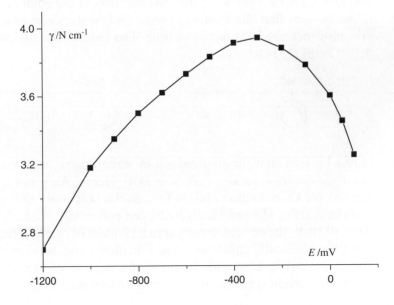

(a) The potential of zero charge is ~ 300 mV negative to that in aqueous NaF. This would imply that the iodide ion adsorbs quite strongly on the mercury surface. The shape of the curve gives no evidence for adsorption of neutral species.

(b) Estimates of the slopes of the graph at 0 and -800 mV lead to surface charges of $+12$ and 6.25 mC cm^{-2}, respectively.

15. From the first table, it can be seen that I_L vs $\omega^{1/2}$ is linear, confirming that the reduction of nitrobenzene becomes mass transport controlled. The data in the second table allow analysis of the shape of the wave. A plot of E vs $\log[(I_L - I)/I]$ is linear with a slope of 60 mV; this is compatible only with a reversible $1e^-$ reaction. Hence, the reduction of nitrobenzene:

Knowing $n = 1$, it is now possible to evaluate the diffusion coefficient from the slope of the I_L vs $\omega^{1/2}$ plot. The slope of this plot is $6.6\ \mu A\ \text{rpm}^{-1/2}$. Using the Levich equation, (7.10):

$$\frac{6.6 \times 10^{-6}}{\left(\dfrac{2\pi}{60}\right)^{1/2} \times 0.12} = 0.62 \times 96500 \times D^{2/3} \times \left(3.7 \times 10^{-3}\right)^{-1/6} \times 2 \times 10^{-6}$$

and $D = 1.2 \times 10^{-5}$ cm^2 s^{-1}. After the addition of the ethanoic acid, it can be seen that the limiting current and/or the slope of a I_L vs $\omega^{1/2}$ plot increase by a factor of four. The reduction now involves $4e^-$. The likely reaction is:

16. It can be seen that the first reduction wave for oxygen at a gold electrode occurs around $+0.25$ V *vs* SCE. Hence, Au is not a good catalyst for O_2 reduction and, in fact, in the potential range of the voltammogram, the product is hydrogen peroxide.

At $+400$ mV, the current density is independent of rotation rate, *i.e.* it is fully kinetically controlled. The potential range between $+350$ and $+200$ mV corresponds to the rising portion of the response where the current depends on rotation rate to some extent, *i.e.* there is mixed control. The pure kinetic current in this potential range is obtained by plotting $1/j$ vs $1/\omega^{1/2}$ and extrapolating to $1/\omega^{1/2} = 0$ to obtain $1/j_{kinetic}$; when $\omega = \infty$, mass transport has an infinite rate and electron transfer must be the only rate-determining step. Such plots are shown below. Log $j_{kinetic}$ vs E is also plotted below and it can be seen that the relationship is linear with a slope of $(120\,mV)^{-1}$. This procedure allows the Tafel relationship to be extended through the mixed controlled potential range.

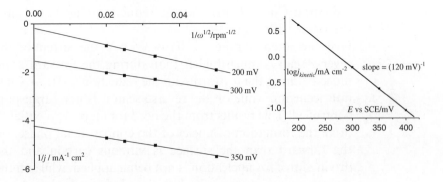

17. Looking at (c), the scan negative from $+400$ mV shows two reduction waves of equal height. Since the electroactive species is Cu(II), this suggests that the reduction occurs in two $1e^-$ steps, Cu(II) $\rightarrow$ Cu(I) $\rightarrow$ Cu, as expected for a chloride medium where Cu(I) is stabilized.

It can be seen from (a) that the limiting current density for the first wave is proportional to the square root of the rotation rate, *i.e.* the reaction is mass transport controlled. The diffusion coefficient for Cu(I) can be estimated using the Levich equation, (7.10). Also, although it would difficult to take the data from the figure, the shape of the wave shows that the Cu(II)/Cu(I) couple is reversible (a plot of $\log[(I_L - I)/I]$ vs E has a slope of 60 mV). For a simple, reversible electron transfer reaction, the forward and reverse steady state scans should be identical. It can be seen in (b) that the hysteresis is minimal and probably results from the scan rate being slightly high for the surface concentrations to be in complete equilibrium; note, however, that there is no anodic current as the Cu(I) is swept away by the rotation of the RDE. The response (c) has several features:

(i) Despite rotation of the electrode, a large anodic peak is observed – this results from the Cu being deposited on the electrode surface. The Cu is oxidized to Cu(I) as this is the stable oxidation state in the potential range of the peak.

(ii) The anodic peak has a symmetrical shape and oxidation ceases positive to the peak when all the deposited Cu is oxidized.

(iii) Positive to this peak, the current density returns to the limiting value for the reduction of Cu(II) → Cu(I); the potential is still negative to the first reduction wave where this reaction is mass transport controlled.

(iv) Indeed, in the potential range of the anodic peak, two reactions occur simultaneously at the electrode surface, the oxidation Cu → Cu(I) and the reduction Cu(II) → Cu(I), the latter with a rate $-j_L$.

(v) Over the potential range −220 to −380 mV, the current on the reverse scan is much higher than during the forward scan; indeed, during the forward scan the reduction Cu(I) → Cu is not occurring while on the reverse scan it is mass transport controlled! This results from the need for an overpotential to drive the formation of nuclei of the copper metal phase; on the forward scan the surface is vitreous carbon and the driving force for nucleation is not being applied, while on the reverse scan a copper layer has already been deposited and nucleation is not an issue.

(vi) The need for nucleation also influences the shape of the Cu(I) → Cu wave – the initial portion shows a very steep rise because nuclei are being formed and their surface areas are growing rapidly because there is a large overpotential for the Cu(I) → Cu reaction at a copper surface (see the reverse scan). Three factors contribute to the steep increase in current with potential (a) an increase in number of nuclei (b) an increase in the size of each nucleus (c) an increase in overpotential.

(vii) The charge for copper deposition and copper dissolution are equal – remember that the potential axis is also a time axis and therefore charge is an area under the curve. But the charges must be measured correctly. The charge for deposition corresponds to when the current is above the limiting current density for the formation of Cu(I) – hence it is the area above this limiting current from −380 to −600 mV on the forward scan and from −600 to −200 mV on the back scan. Some of the area gets counted twice as the current is passing on both forward and back scans. The anodic charge is the area under the symmetrical peak calculated from the cathodic limiting current for the reduction Cu(II) → Cu(I) – see (iv) above.

18. Cyclic voltammograms at stationary and rotating electrodes for reactions involving reactant and product in solution differ in two critical ways:
    (i) At a stationary electrode, the response for each reaction is an asymmetric peak because mass transfer occurs by non-steady state diffusion (Section 7.3.2.1). At the RDE, the voltammogram has a sigmoidal shape, see previous question.
    (ii) At a stationary electrode, the product(s) of the electrode reaction are not swept away from the surface so that they are available for reaction on the reverse sweep.

With the Cu(II) solution, the voltammogram shows a reduction peak for the reaction Cu(II) → Cu(I) on the forward scan and an oxidation peak for the reaction Cu(I) → Cu(II) on the back scan. The peak separation is 60 mV, as expected for a reversible 1e⁻ reaction. Variation of the potential scan rate would show that $j_p \propto v^{1/2}$ and the diffusion coefficient could be estimated using Equation (7.53).

19. The experiment with the ferrocyanide solution allows the determination of the collection efficiency; taking the ratio of the currents when the oxidation of ferrocyanide is mass transport controlled, $N = 0.22$.
The voltammogram for the reduction of oxygen at vitreous carbon shows a slightly peaked wave at $E_{1/2} = -0.36$ V and the ring response (ring held at a potential where hydrogen peroxide oxidation is mass transport controlled) follows this shape. Hydrogen peroxide is formed at the disc since if oxygen were reduced to water there would be no ring response. The ratio of the ring to disc currents at the peaks is ~ 0.2 – this compares with the collection efficiency so almost all the current at the disc leads to the formation of hydrogen peroxide.

20. Since the product of the corrosion of copper on open circuit is detected on the ring, it must be soluble in the chloride medium. The only possible species is Cu(I) or Cu(II) and, since it is oxidized on the ring, it must be Cu(I); this also coincides with our knowledge of copper chemistry in chloride media; Cu(I) is strongly stabilized. Corrosion stops in the absence of oxygen and hence the reactions on the disc are:

$$Cu - e^- \longrightarrow Cu(I)$$

$$O_2 + 4H^+ + 4e^- \longrightarrow 2H_2O$$

On open circuit, there is no net current flow so that the electron fluxes from the two reactions are equal. At the ring, the reaction is:

$$Cu(I) - e^- \longrightarrow Cu(II)$$

From the ring current and the collection efficiency, the partial current for the formation of Cu(I) is:

$$I = \frac{26}{0.34} = 76.5\,\mu A$$

The concentration of Cu(I) after 30 min can then be found from a Faraday's law calculation:

$$[Cu(I)] = \frac{76.5 \times 10^{-6} \times 1800}{96500} \, 63.5 \times \frac{1}{100} = 0.91 \, \mu g \, cm^{-3}$$

This is very close to the concentration found by atomic absorption spectroscopy.

21. The plot of I vs Q is linear and extrapolates to 600 C for complete reduction of the 145 mg of 2-fluoronitrobenzene. Using Faraday's law:

$$\frac{600}{n \times 96500} = \frac{145 \times 10^{-3}}{141} \, \text{giving } n = 6$$

The j vs t data give a linear $-j$ vs $t^{-1/2}$ plot with a slope of $1.03\,mA\,cm^{-2}\,s^{1/2}$. Hence, using the Cottrell equation, (7.36), $D = 6.9 \times 10^{-6}\,cm^2\,s^{-1}$.

22. The voltammogram recorded at $1\,V\,s^{-1}$ shows the response for a reversible 1e$^-$ process centred around $-0.98\,V$ with a small distortion at more negative potentials (most obviously a bump on the reverse scan). The response at $0.1\,V\,s^{-1}$ is quite different. The current densities are $\sim 10^{1/2}$ smaller and this is because of the longer timescale of the experiment causing the rate of non-steady state diffusion to have decayed further. More interestingly, the anodic peak at $-0.95\,V$ associated with the reversible 1e$^-$ process has diminished and a new reversible peak is becoming obvious at more

negative potentials. The voltammograms are consistent with the mechanism:

In fact, the voltammograms lead to a rate constant for the cleavage of the C–Br bond of $0.4\,s^{-1}$. The mechanism could be confirmed by recording more scan rates and/or an electrolysis at $-1.05\,V$ that should lead to nitrobenzene as a major product.

23. Plotting I vs Q and extrapolating to zero current gives a charge for the complete consumption of A equal to ~ 1020 C; using Faraday's law, this give $n \approx 2$. Also, the voltammogram with a positive limit of $+1.4\,V$ suggests that the first step is a reversible $1e^-$ oxidation leading to an aromatic cation radical. Further $1e^-$ oxidation of this cation radical leads to an intermediate that is consumed in a rapid chemical reaction. Notably, the existence of this second step close in potential to the first leads to some distortion of the cathodic peak on the back scan. Hence, the voltammetry is consistent with the proposed reaction. Confirmation would require an electrolysis at $+1.6\,V$ and identification of the products using chromatography and/or spectroscopy.

24. The voltammograms show both anodic and cathodic peaks; both are symmetrical (suggesting that the electrochemistry involves a change in a surface layer) and the areas under the peaks look similar. These observations, together with the fact that the voltammograms can be repeated without cleaning the electrode, indicate that the surface is in the same state at the beginning and end of each potential cycle; the changes are chemically

reversible, if not electrochemically reversible (there is a clear peak separation). Analysing the voltammograms more quantitatively:

(i) the peak current densities are proportional to the potential scan rate;

(ii) the charges associated with the anodic and cathodic peaks are very similar;

(iii) at all scan rates, the charges associated with the anodic peaks are $\sim 0.8\,mC\,cm^{-2}$. Normally, the $1e^-$ oxidation/reduction of a monolayer at an atomically smooth surface requires 0.1–$0.2\,mC\,cm^{-2}$. Hence, with Ni in base, the chemical change involves either a monolayer over a slightly rough surface or a few monolayers on a rather smoother surface. The chemical change is normally written:

$$Ni(OH)_2 + OH^- - e^- \rightleftharpoons NiOOH + H_2O$$

25. As in the last question, a symmetrical anodic peak is observed but there is no corresponding cathodic peak; the chemical change is completely irreversible and it changes the surface since subsequent cycles are strongly affected. Moreover, on the reverse scan, there is significant anodic current (oxidation continues to occur) and the anodic current even increases again in the potential region of the peak and continues to more negative potentials. Notably, the total anodic charge passed during the voltammogram is $\sim 80\,mC\,cm^{-2}$ and this corresponds to massive corrosion to a solution soluble species or a very thick layer corrosion film.

It is likely that the initial anodic current corresponds to active Ni corrosion and the peak arises because, positive to the peak, passivation occurs, resulting from deposition of a poorly conducting corrosion film on the surface of the nickel. However, the film continues to thicken during the scan to positive potentials from -0.4 to $+0.4\,V$ and also during the reverse scan. When the potential reaches the 'active corrosion' range the anodic current increases again. During the potential range -0.5 to $-0.75\,V$, the current is much higher on the reverse scan than the forward one and hence the surface must be more 'active' for corrosion. The most likely explanation requires the recognition that we always measure a net current (from both cathodic and anodic contributions to the measured current) and hydrogen evolution (a cathodic process) may be occurring on the initial surface. Indeed, at the corrosion potential of the polished nickel, *ca.* $-0.7\,V$, the currents for nickel dissolution and H_2 evolution are equal. The corrosion potential moves negative after the first scan. This could arise if hydrogen evolution is strongly inhibited by the corrosion film. A consequence would be that the measured current would be anodic over the potential range where H_2 evolution is inhibited.

26. The corrosion current density can be calculated from the metal loss rate using Faraday's law (noting 1 year $\approx 3.15 \times 10^7$ s):

$$\frac{j_{corrosion} \times 3.15 \times 10^7}{2 \times 96500} = \frac{0.3 \times 10^{-3}}{56} \text{ and } j_{corrosion} = 0.033 \text{ μA cm}^{-2}$$

This is the partial current density for iron dissolution at the corrosion (open circuit) potential, -510 mV. Since no net current density can flow at the open circuit potential, this is also the partial current density for H_2 evolution at this potential. Since the Tafel slopes for the two reactions are known, the current densities for each reaction at -825 mV can be calculated:

$$\text{For Fe dissolution, } j = 3.3 \times 10^{-8} \exp\frac{-825 + 510}{26}$$

$$= 1.85 \times 10^{-13} \text{ A cm}^{-2}$$

$$\text{For } H_2 \text{ evolution, } j = 3.3 \times 10^{-8} \exp -\frac{-825 + 510}{52} = 14.1 \text{ μA cm}^{-2}$$

The rate of iron dissolution has decreased by a factor of over 1 million to 1.7 ng cm^{-2} year^{-1}. There are two consequences: (a) H_2 is being evolved, although only at a rate of ~ 50 cm^3 cm^{-2} year^{-1} and (b) the zinc anode is dissolving. Indeed, the partial current density for zinc dissolution at -825 mV must be 14.1 μA cm^{-2} as no net current can be flowing at this zinc protected steel structure; hence the loss rate is ~ 140 mg cm^{-2} year^{-1}. The rate of loss of zinc from the sacrificial zinc anode is substantially higher than from the unprotected steel surface but the corrosion is localized to the zinc anodes and these can be replaced periodically.

Subject Index